Once Upon a Forest

Bibliothèque et Archives nationales du Québec and Library and Archives Canada Cataloguing in Publication Data

Main entry under title:

Once upon a forest

Translation of: Il était une fois la forêt.

1. Forests and forestry - Québec (Province). 2. Forest conservation - Québec (Province). 3. Forest degradation - Québec (Province). 4. Lumbering - Québec (Province). I. Quentin, Isabelle, 1955- . II. Mangin, Patrice J.

SD146.Q4I413 2010 333.7509714 C2010-941966-9

Legal deposit
Bibliothèque nationale du Québec
National Library of Canada
Printed in Canada

Distributed in Canada by:
Prologue
1650 Lionel-Bertrand Boulevard
Boisbriand, QC J7H 1N7
Canada
1-800-363-2864
www.prologue.ca

Distributed in the USA by:
Enfield Publishing & Distribution Co.
234 May Street
Enfield
New Hampshire 03748
United States
603-632-7377
http://www.enfieldbooks.com

Interview recording: Claire Demers
Translation: Gary Rosenberg
Translation and revision: Keren Penney
Graphic design and page layout: Manon É. Léveillé
Processing of illustrations and photographs: Quentin Photographe (Évelyne Deshaies)
Printing: Transcontinental

http://iqe.qc.ca
ISBN: 978-2-922417-83-8

Mixed Sources
Product group from well-managed forests, controlled sources and recycled wood or fiber
www.fsc.org Cert no. SW-COC-000952
© 1996 Forest Stewardship Council
FSC

1 2 3 4 5 12 11 10

Once Upon a Forest

Remarks collected by Isabelle Quentin,
under the direction of Patrice Mangin.

iQ

Table of contents

The forest as landscape

WHAT IS LANDSCAPE? Landscape is born of the ability we have to appreciate various aspects of territory through the interaction of body and place. It is an experience in relation to land. Landscape is a social construction; it is an invention that does not exist in and of itself.

One of the first important forms of landscape was the mountain. In Europe, during the Age of Enlightenment, the writers and painters of England who first journeyed to Italy had to cross the Alps. Inspired, these travellers portrayed the mountains as something beautiful and sublime, and transformed people's perceptions of them. Then there was the sea. When people began to venture out of the cities for the fresh air of the seaside, they realized that the oceans and beaches were more than places

Philippe, talk to us about landscape...

where shipwrecks occurred, they were places for contemplation. They saw the sea depicted in paintings and photographs, and experienced it in a positive way.

The notion of forest evolved a little differently. A long time ago, the forest was first invented in myths and fables, through imagination. Even today, when we allude to the forest, something stirs deep within us. That is because the forest incarnates the deepest essence of nature, even more powerfully than the mountains or the sea.

The forest has even contributed to the representation of our country, its image. The Canadian identity is affirmed through landscape, as expressed, for example, in paintings by the Group of Seven in the 1920s. Through the celebration of nature and the forest, landscape has thus contributed to forging or strengthening the Canadian territorial identity.

To sum up, then, seen from these various perspectives, landscape is a social and cultural invention, conceived and communicated through the way we see territories and places.

A JOURNEY THROUGH THE FORESTED LANDSCAPE In the Bas-Saint-Laurent region, the forest is literally at the end of the road. If you start at the river, you can clearly see the hand of man extending along its shores. One village, two villages... the forest was uprooted again and again, until, one day, man got tired and said: "We'll leave the woods alone now; we'll return later."

If you explore the back roads, you notice they are arranged in a series of terraces. You get a really clear sense that there is a story behind the creation of this territory. Human habitation was developed in relation to the topography and the circumstances of the place. Further on, beyond the back roads, you find the woods.

Wherever the land was cultivated for agriculture, whenever the soil was worked, stones came to the surface. They were picked up and heaped into piles, usually around groups of trees, creating kinds of islands in the middle of huge fields. Over time, they became inseparable from their location, a part of the landscape.

To get to the forest, you have to follow a forest path. But you cannot enter the forest directly. There is an area where the trees have been cut down, and the first thing we come to is an edge that has been cleared along with some rather chaotic terrain. Discovering this forest is, in a way, discovering chaos.

So you have to walk for kilometres along a boring road. It is boring because there is not much to see. Everything that might have had some value has been removed, cut down. All that is left, really, is an abandoned road. On one side, stumps and the occasional sparse row of conifers, nothing more.

The further you go, the more you notice some species making a comeback. Evidently, after a clear-cut, it is not easy for the trees to grow back. We are still not in the forest, but rather in a place that might have been a forest at one time.

Depending on the season, this forest reveals different aspects of itself. Right now, it is early spring, a time for discovering texture. Forest paths often have areas of higher elevation that serve as lookout points. You can see a velvety patchwork of texture now that would not have been visible a month earlier. Pines are distinguished from their leafy cousins in brilliant bursts of delicate green or yellow. All these nuances will disappear in a month, and other textures will take their place. You realize that the forest landscape is a landscape of expression, not merely of land, as it may appear at first glance, but also of seasons, as conveyed through this mass of trees. So many textures—a horizon of texture—and such depth... It is very poetic.

These partly devastated landscapes are very interesting because they highlight various features. Like these birch trees that were of no interest to the woodcutters, and which now occupy a significant amount of space. Or these panoramic viewpoints that would not exist in a forest left to itself.

Here, as you approach the edge of the actual forest, you hear the rustling of leaves in the wind and begin to anticipate a more sensory experience. Landscape is not only

what we see, it is
also what we hear, smell,
and touch.

IN THE HEART OF THE FOREST The forest has a heart, and it is fascinating. You do not want to enter it without a compass, because after only 20 or 30 meters, you may find yourself completely lost and disoriented, even at noon on a sunny day! From the moment you set foot in the forest, you are enveloped by both sound and silence. You become immersed in a universe that has you completely captivated, where the slightest sound can make you jump, and silence can frighten you.

Even though it is difficult to move forward in this kind of endless mass, you feel you could keep on walking forever, all the way to infinity. You make contact with the trees in a way that is quite disconcerting. You bump into trunks, walk into branches, trip over stumps... It is as if nature were suddenly challenging you. It is a very active experience! A hundred meters like this, and you have worked up a sweat. You run, you get covered in scratches, you cannot get out unscathed. And the further you go, of course, the more mosquitoes and black flies there are. But in spite of that, it is a thrilling experience!

You become mesmerized by the trees, dazzled by the light filtering down through the leaves. As you continue on your journey, you discover brooks and a host of other wonders. It is as though the forest were hiding something, an inner landscape that never ceases to amaze.

In the heart of the forest, you lift your eyes. You look to the sky. And you discover a fantastic tableau of shapes, because, here, the sky acts like a backdrop. This abundant nature feeds our imagination; the day it no longer nourishes

us and awakens our senses is the day when we will have lost something precious. It is incumbent on us to preserve and, above all, cherish it.

AN UNEXPECTED CHROMATIC AND SENSORY UNIVERSE In terms of landscape, the forest is riveting, both in the way it regenerates itself and in the way it dies and decays.

Look over here, at these Halloween-like images: trees covered with moss, lichens growing on trunks and branches.

The contrasts are fascinating. The interior of the forest actually has a wide chromatic range. People tend to think of the forest as only green. In fact, the forest boasts a multitude of colours, a veritable palette of distinct shades and textures.

Over there, the trees have split apart, like this almost centenarian, which now reveals its rough, woody matter within.

And there is a lake made by beavers. Part of the forest has died because of the actions of an animal that built its habitat and transformed the ecosystem, and this has created an entirely different universe for us.

Suddenly, we come upon a space where silence reigns. Not a breath of wind. A huge windstorm has already levelled everything.

Further on, there is a tree that was spared by the lumberjacks. It is a white pine, a remarkable tree. It is a vestige of what the forest was like about two or three hundred years ago. This pine is at least 30 meters tall. It towers high above the present-day forest, which barely reaches 15 meters. It stands like a venerable elder statesman in this territory, and provides us with a glimpse of what the forest must once have been. It is the witness of a distant past, a time when its species was highly valued, when it was an integral part of Canada's economy. Oaks and white pines were, in effect, the two noble tree species that were brought back to Europe. And this one is still here!

Overall, it is pretty quiet in the forest. The loudest things here are the squirrels. But when they move around, you would swear a moose was coming towards you. It is impossible for a squirrel to make as much noise as a moose with its great big hooves, of course, but in the forest, sound can become totally distorted. Such is the effect of immersion, of the state of mind the forest plunges us into. Talk to people who gather things or hunt. They will tell you that after a few hours in the forest, something happens...

The climate that sculpts the forest

Dominique talk to us about climate...

THE ARTIST AT WORK In Canada, we perceive the passing of the seasons with our eyes as much as with our bodies. The changing beauty of the forest is an expression of seasonal transitions. Climate variations offer us landscapes that are by turns dense, bare, verdant, tinged with red, and carpeted with white, pierced by conifers that do not lose their needles. These annual changes influence how we perceive the forest. And this changing palette brings tremendous quality to our life.

Climate is the great sculptor of the forest. In ecology, heat and water are the two most important factors that influence life. And the forest is full of life! Climates vary, and forests—tropical, temperate, or boreal—adapt. The main characteristics of the forest are, to a great extent, determined by climate, and especially by temperature and moisture.

Moving from south to north, the forest changes. There is less and less heat and water, and a greater variation between the seasons. In warm and humid climates, trees grow tall and generally have large leaves. A wide variety of different life forms thrive there. In colder climates, growth is more arduous, slower. The forest is not as dense, and there is less variety. Just by looking at a photograph of the forest, you can tell the climate of a region.

The more hours there are in a year with favourable temperatures, the more plants grow and flourish. The number of degree-days—the sum of the temperatures necessary over a given period to allow the growth of vegetation—varies greatly from one region to another. Because plants are largely made of water, an abundance of water also favours growth.

Tree shapes tell us about a forest's productivity and vitality. If the forest is full of deciduous trees, with relatively small, but supple and tender leaves, then we are in a temperate forest. We can deduce that the climate is seasonal, that there are winters, but that there is also enough heat in the summer for the forest to grow well.

If we come to a forest where the trees are smaller and have needles, we know that the climate is harsh, with vastly different summers and winters, and that either the cold is fairly intense or there are long periods of drought. And if conifers are the only species, it means that it is cold for a good part of the year.

The finer, more subtle characteristics of the forest are determined by a host of elements, including snow, frost, wind, temperature stress… all of which are part of climate. All these factors go into sculpting the forest's details. For example, in certain places, wind can cause huge trees to fall, thus influencing the renewal and propagation process of the forest. Where it is frequently strong, wind will cause trees to shrivel and curl inward. Frost sometimes causes the cells of plants to burst when the water they contain expands. Some plants are able to resist this, others not. And an accumulation of snow can protect certain species of plants and animals against intense cold.

Thousands of little brush strokes and chisel marks, everywhere, all the time.

CLIMATE CHANGES We often disassociate climate from the forest, as if there were no direct link between what happens in the air and what happens on the ground. But everything is related...

The current global warming, for example, favours our forests to some degree. Trees grow more when it is warmer, and a greater number of species are able to flourish. But because this is happening so quickly, the forest will not necessarily be more beautiful. There will be a period when the best adapted species will not have arrived yet: they will still be too far south and will not have had time to colonize more northern countries. Or insects could arrive faster than the trees and eventually destroy the forest. This would be a chaotic development we would be partly responsible for.

Consider the pine beetle infestation that Western Canada is now having to cope with. In the past, this insect was not able to reproduce in sufficient numbers during summer, but today it devastates entire areas of forest. In the long term, the climate will be warmer and more humid, which will favour the northern forest... and cause dramatic changes to its makeup!

We know, from having reduced acid rain and shrunk the hole in the ozone layer, that we can reverse certain trends when we implement measures like those contained in the Montreal Protocol. But in the case of climate change, the problem is more complex.

THE FOREST ITSELF SCULPTS CLIMATE Although we have known for a long time that climate influences the landscape, it is only recently that we have understood that the reverse is also true. In some ways, the forest actually sculpts climate.

For example, if, in Quebec, we removed all the forests, the climate would be much drier. That is because the forest acts like a sponge. When water falls, it falls on the ground, on leaves, and on plants, before warming and evaporating. If there were no forests, water would run directly into the rivers, and, eventually, out to sea. If you look at a photograph where there is a forest on one side and none on the other, you can clearly see clouds above the forested area. You can also feel a difference in the air when you approach a wooded area in the city: the climate changes suddenly and becomes cooler and more humid.

This is how it works: the forest affects humidity, influences the reflection of heat into the atmosphere, and captures carbon, through a process called photosynthesis. This is particularly true of the

boreal forest, which covers 70% of Quebec and a very large part of Canada. Plants absorb carbon dioxide and, using solar energy, produce cellulose. The resulting waste material is oxygen! For us, oxygen is an essential element, but for plants it is just residue. The cellulose they produce forms trees, whose leaves and trunks eventually fall. In a tropical forest, everything decomposes very rapidly, while in a northern boreal forest, plant matter accumulates on the ground, stocking carbon. In other words, the boreal forest removes carbon dioxide from the atmosphere and retains it for long periods, which helps clean the air we breathe.

The Canadian forest, taken as a whole, constitutes a large portion of the world's forests. It plays an important role in the atmosphere of the entire planet, in its composition, and in the great continental exchanges of air masses.

THE EFFECT OF CLIMATE ON PEOPLE THROUGH THE FOREST Human beings are better equipped than forests to adapt to rapid climate change. As animals, we can regulate our body temperature. We know how to make fire. We can clothe ourselves. But our lives are still very dependent on our environment.

To illustrate this point, let us look at Western Canada, where the forests are exceptional. At the time of the arrival of the white settlers, life for the Native people was easier on the coast of British Columbia than in the Yukon. The climate was milder, salmon was abundant, and the resources of the bountiful forest lay at their doorstep. They had more time to develop their intellect and their art compared to their Yukon neighbours, who were constantly searching for food, who had to travel long distances to hunt, and who were never certain of the future.

Biodiversity and the forest

Daniel, talk to us about the forest and its biodiversity...

AN ASTRONOMICAL DIVERSITY ON EARTH! How many living species are there on the planet? At the present time, a little under two million have been identified. But those are only the ones we know about. The most conservative estimates place the number of species on earth at ten million. That means there are still many left to discover. And this does not take into account that, in some regions of the globe, species are disappearing even before we are aware of their existence.

Most of our knowledge is centred around birds and mammals, because we are more curious about what resembles us. In general, people have little, or no, interest in studying molluscs, animals that are not attractive, or those that inspire fear. They hold less appeal for us.

Yet, we are amazed when we learn that in Madagascar—an island that has lost 90% of its forest—we are still finding new species of mammals, more precisely lemurs, the most primitive group of primates, which exist nowhere else. In fact, on large oceanic islands, the rate of endemism (the presence of unique species) is very high.

What we are less aware of is that even though the rate of endemism is relatively low in Canada—given that it is part of a vast continent—it is also home to unique species. And it is our responsibility to ensure their conservation, because no one is going to do it for us.

Did you know, for example, that the population of a certain fish species in the Richelieu River, the copper redhorse, has dropped to approximately 300 individuals, and that this fish is not found anywhere else in the world?

HUMAN PRESSURE ON NATURE Environmental problems are caused by an excess of human population, when too many people use too many resources. There are many people on earth, and there are going to be more and more. Most experts agree

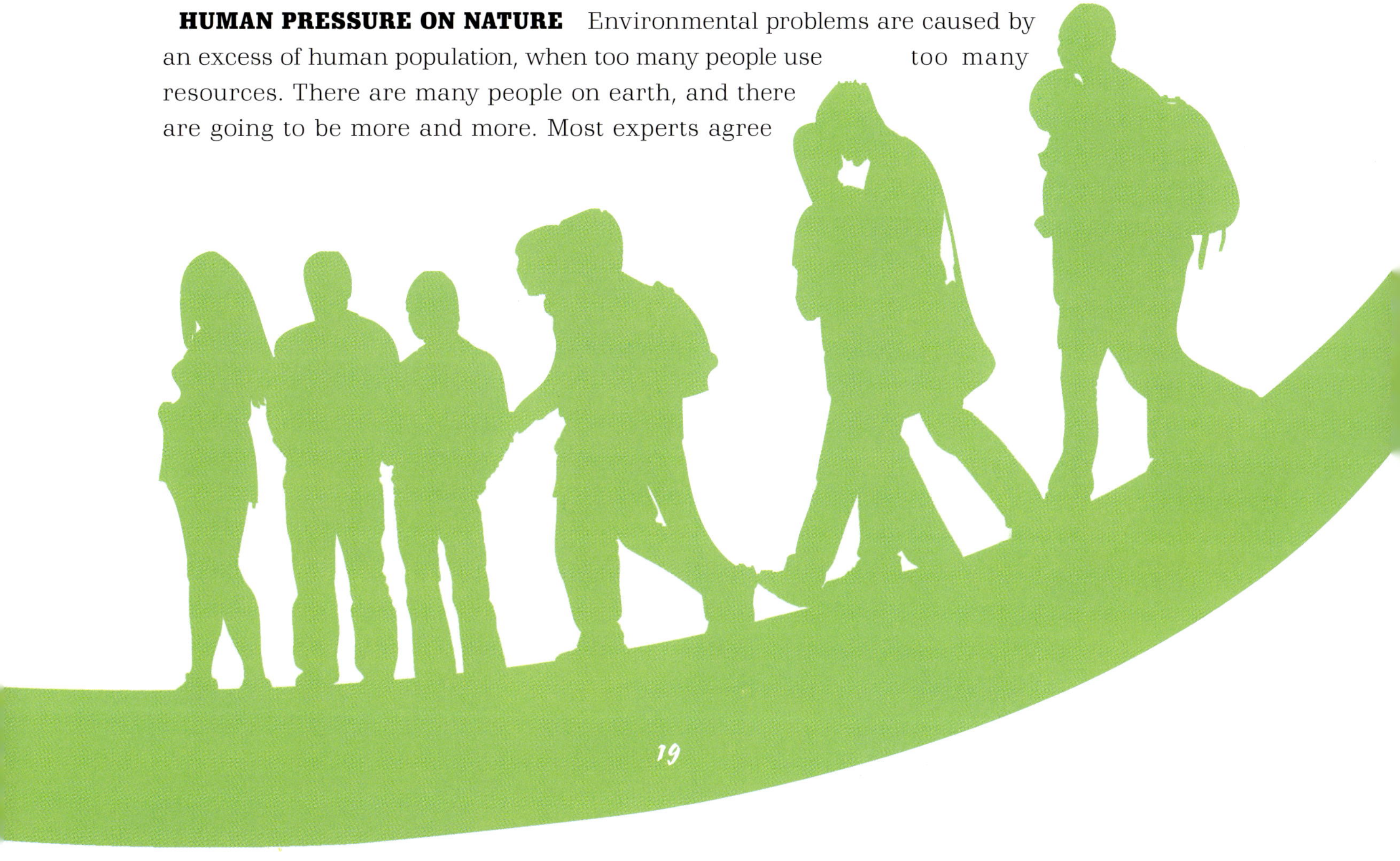

that we are not far from reaching the maximum number of human beings the planet can feed. And still, the world's population continues to grow.

Wealth, as we know, is distributed very unevenly. Resources too. All the arable land on the planet has already been exploited. If we succeed in being able to produce more food, it is because we do it in a more productive way, not because we increase the land's surface. That said, 40% of the living matter produced by the planet is already been exploited by humans—one species, it must be remembered, among millions of others!

Biodiversity is under pressure. It is threatened by the loss of habitat, overexploitation, and the introduction of exotic species—a result of the acceleration in world trade.

BIODIVERSITY: INTERCONNECTED STORIES

Our deciduous forest Fifty million years ago, in the Tertiary age, the Northern Hemisphere was dominated by deciduous forest. The present-day continents of North America and Eurasia formed a single mass. As a result, the forest was continuous.

Then the continents drifted apart, and mountain ranges formed. The one that concerns us here is the Rockies, young mountains like the Himalayas. So what do large mountain ranges do? Well, they cut the dominant winds in one region, creating patterns of precipitation we call orogenic, or caused by mountains. Mountains are created by the drifting of continents. That means that the mountains were not always there. When the Rockies appeared, they created a huge dry zone to the east: the Prairies.

The zones of vegetation of deciduous forests with humid climates, like those found in western Europe, eastern Asia, and eastern North America, are what remain of those upheavals. With this in mind, we can better understand how similar patterns of biodiversity can be seen in the forests of eastern North America and eastern Asia. And why, for example, we find ginseng in both Asia and North America.

Ginseng In 1716, Father Jartoux, a Jesuit, was working for the emperor of China. Back then, the Jesuits were the scientists of the Western world, and Father Jartoux was a specialist in surveying and cartography. He learned that the emperor valued ginseng enormously and bought roots for three times their weight in silver. After reading the correspondence from his American colleagues in *Jesuit Relations*, where each related what he had seen in the world, Jartoux wrote that, in China, there were many species similar to those found here: maples, oaks, beeches... and the highly prized ginseng, which he made a drawing of.

Back in Montreal, Father Lafiteau could not find any ginseng, so he showed the illustration to the Iroquois, who were very familiar with the famous root and used it as a medicinal plant. Its Latin name, *panax*, comes from the Greek "panacea." The plant was said to treat everything because it had an almost human form, with appendages that looked like arms and legs. The Jesuits began exporting ginseng to China through the French India Company. At the time, it was the second largest export after furs! And so began the rush for ginseng, which peaked, then just as quickly declined, all in less than 40 years!

Father Marie-Victorin reported that instead of letting the roots dry slowly in a loft, they dried them in ovens to speed up the process. Its popularity resulted in the disappearance of Canadian ginseng. Soon, all that remained was a proverb among people when speaking of something that failed: "C'est tombé comme le ginseng" (It fell like ginseng). The ginseng hunt—which practically eradicated the plant from the forests—has today been replaced by its cultivation under artificial shade, largely in Ontario.

It takes two years to produce a ginseng root under artificial shade, compared to the 15, 20, and sometimes even 35 years it would take to grow in the forest. Like wild leek, ginseng is a valued forest product that is not wood. Unfortunately, picking wild leek or ginseng means removing the entire plant. When harvesting leads to the death of a plant, it becomes important to decide what should and should not be gathered in the forest. Incidentally, selling wild leek is prohibited in Quebec, because it was disappearing from our forests under the pressure of being harvested for commercial purposes.

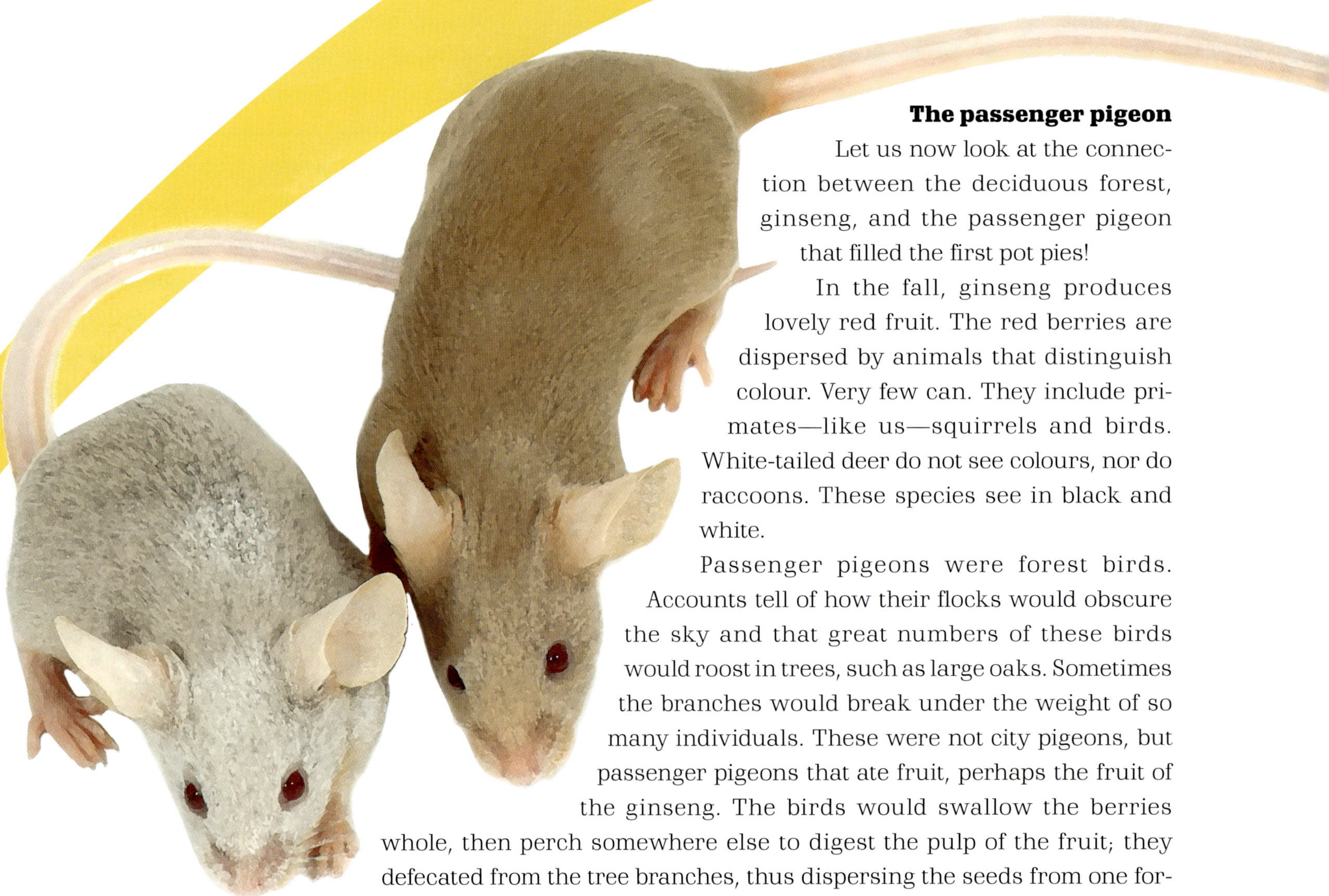

The passenger pigeon

Let us now look at the connection between the deciduous forest, ginseng, and the passenger pigeon that filled the first pot pies!

In the fall, ginseng produces lovely red fruit. The red berries are dispersed by animals that distinguish colour. Very few can. They include primates—like us—squirrels and birds. White-tailed deer do not see colours, nor do raccoons. These species see in black and white.

Passenger pigeons were forest birds. Accounts tell of how their flocks would obscure the sky and that great numbers of these birds would roost in trees, such as large oaks. Sometimes the branches would break under the weight of so many individuals. These were not city pigeons, but passenger pigeons that ate fruit, perhaps the fruit of the ginseng. The birds would swallow the berries whole, then perch somewhere else to digest the pulp of the fruit; they defecated from the tree branches, thus dispersing the seeds from one forest to another. Today, birds do not eat and disperse these berries anymore. It is more common to see small rodents feeding on them, and they usually bite and eat the seeds.

Nowadays, ginseng can be found in the wooded areas of southern Quebec, isolated amidst seas of corn and soybean. A few vestigial populations have survived. If they disappear from a woodland area, one might hope that they would be reintroduced by a bird dispersing its seeds. But the reality is that, today, ginseng seeds are only dispersed by gravity—which means that they fall right beside the plant... and that does not get them very far!

All this to illustrate that we do not fully understand the real value of the passenger pigeon we lost… and not just as an ingredient in a meat pie! It may have played an important ecological role. Each species has its role in nature, and we are not always aware of it. In truth, we know very little about the intrinsic value of most organisms.

LIVE AND LET LIVE We often hear that we have to enhance the value of our forests, as though, if we do nothing, the forest has no value at all. However, this way of thinking can be qualified. Some, like the Aboriginal peoples, say that the forest has been bestowed upon us, and other people elsewhere in the world believe that we must manage the forest's species for our children.

The human race is evolving. Today, we find it unacceptable to reduce other human beings to slavery or, worse, to kill them. Maybe we will soon apply this way of thinking to other species... Our philosophy has already started to change. Thirty years ago in our national parks, for example, we could see signs like: "Do not pick the flowers; let others admire them." Now, they say something along the lines of: "Do not pick the flowers; they have as much right to exist as you. You are in their habitat, so please do not disturb them." The emphasis is no longer placed on depriving the next visitor of pleasure, but rather on not destroying individuals of a species. That, not other people's enjoyment, is now what is considered important. This is a kind of altruistic view of nature, a belief that we should let other species live, which has evolved in conjunction with our "live and let live" mindset. We are stewards not only of our human neighbours but also of the other species with whom we share this world, the world that belongs to us all—the earth.

The Aboriginals' forest

A FOREST IN THE SKY

When I was little, my grandfather, Paul Sioui, Théophile Gros-Louis, and the other village elders did not say "I am going to the forest." They said "I am going to the Sky." They left the meaning up to our imagination. And that is probably the way all First Nations people see the Sky. Like a forest, a tranquil, happy place of peace, harmony, and balance, where everything has its place, where human beings cannot exist other than in perfect synergy with what is around them, in contemplation of their own condition...

It is here that we see the vast difference between people with circular thinking, and people with linear thinking—two great philosophies of life, two great principles, and two great adversaries.

Konrad, talk to us about the relationship between the forest and the Aboriginal people...

CIRCULAR THINKING Circular thinking is indigenous, natural thought, where the least thing counts, where everything is in harmony and everything is there to question you. Everyone is equal around the circle, and everyone sticks together. For example, I cannot not leave without my grandmother: she is important to me, she is a part of me.

Nature, the forest, is who I am. I belong to the forest and it belongs to me: we are inseparable. I have an obligation to take care of it because it is my mother, the one who nourishes me. Everything that nourishes is a mother. The same goes for the land, which I also have an obligation to care for.

I am called superstitious, because after eating the beaver or the moose that nature gave me, I do not throw away the bones or the antlers. I hang them in the trees. When we go into the forest, we know whether Huron people have been there, because they have a different way of hanging the remains of animals than the Inuit, the Cree, or other First Nations peoples. Nonetheless, we all hang our bones. We do not throw anything in the fire because, while it may provide some beneficial heat, the spirit of the animal could escape. On the contrary, hanging the bones shows respect, and ensures that the animal's spirit can return at any time. We do not want it to leave, because we need it. If we have taken good care of the animal, if we have used all its parts without wasting anything, and if we hang its bones or its head, it will remain with us. Then, when we or other people return, there will still be game to hunt.

That is what the forest is to us: our church. Each leaf on every tree is our bible. The forest is freedom and peace, and at the same time it is our pantry, our refrigerator, the place where we can always find things to nourish us and thus ensure our survival. But all this has to be done with the greatest respect, because we too are part of the forest we live in. That is what circular thinking means.

LINEAR THINKING In contrast, linear thinking is fundamentally individualistic, violent, and aggressive: the forest belongs to me and I am going to exploit it, I am going to tame it, and even empty it completely, to satisfy my needs. The forest is mine, and I am going to take whatever I want from it to nourish myself and serve my interests.

The forest has suffered as a result of linear thinking, and it continues to suffer because there is nothing to stop those who engage in this form of thinking. If they decide they want to get from point A to point B, they will remove everything that lies in their path, whether it is their father, their mother, their brother, their sister, or anyone or anything else. Animals, rivers, and forests are, in their eyes, only waiting to be dominated and exploited, so that they can achieve their goals.

This way of thinking came from Europe, but it has never taken root among the First Nations people, or any of the other peoples on earth whose thinking is circular: the Maoris of Polynesia, the Aboriginals of Australia, the pygmies of Africa, the Mayans of Central and South America... circular thinkers have never been tamed. Circular thinking goes back many thousands of years, and it is no secret to anyone: we do not want to become linear thinkers. That is our rallying cry, and we believe that the next century will prove us right.

PROTECTING WHAT YOU LOVE The forest, life, is so fragile. Circular thinkers are also fragile, like the animals. A handful of linear thinkers can subdue thousands of circular thinkers. Not because the circular thinkers are naive, but because their way of thinking does not permit aggression, violence, massacre, torture, or murder, and because they are unable to think any other way. According to our philosophy, wars of adoption or hand-to-hand combat are acceptable, but wars of extermination are not.

PROTECTING THOSE YOU LOVE We must remind ourselves that human beings are part of biodiversity.

Without intending to, the French and other Europeans brought with them microbes that devastated entire Indigenous nations. Among the Wendat alone—whom the French called Huron—whose population was over 40,000 at the time of Champlain, the pandemic spread like wildfire.

The French, upon arrival, were also ill. The great Huron-Wendat chief Donnacona warned Jacques Cartier: "Do not continue on, do not go to Hochelaga; in the big river, you will become trapped." But the French were searching for gold. They did not listen, and they got stuck at Deschambault. They fell ill, but had the good fortune to be treated by our ancestors, who brought them medicinal plants from the forest. They knew the properties of all the herbs, of all the plants. And that knowledge has been handed down from mother to daughter ever since.

My mother herself was a great medicine woman, as was her mother before her, who treated people all her life. She looked everywhere for healing herbs. She belonged in the forest, it was her home.

The healing forest

ANCIENT KNOWLEDGE In the beginning of time, men banded together in groups for strength and to be able to track game more effectively, while women stayed behind, especially when they were pregnant and physically weakened. Today we know that women's bodies consist of less muscle and protein than men, but more hormones and life-supporting fat. They stayed close to their dwellings, either caves or huts, and learned about the ecosystem around them to nourish themselves and care for their children. In the south, they discovered that if they kept the seeds of a given species and planted them at the right time, they could cultivate crops like corn or squash. In the north, people were more gatherers and nomads, who followed the cycle of seasons, the fruits, and the herds. If women knew more about plants and herbs, it was because they were closer to their immediate environment.

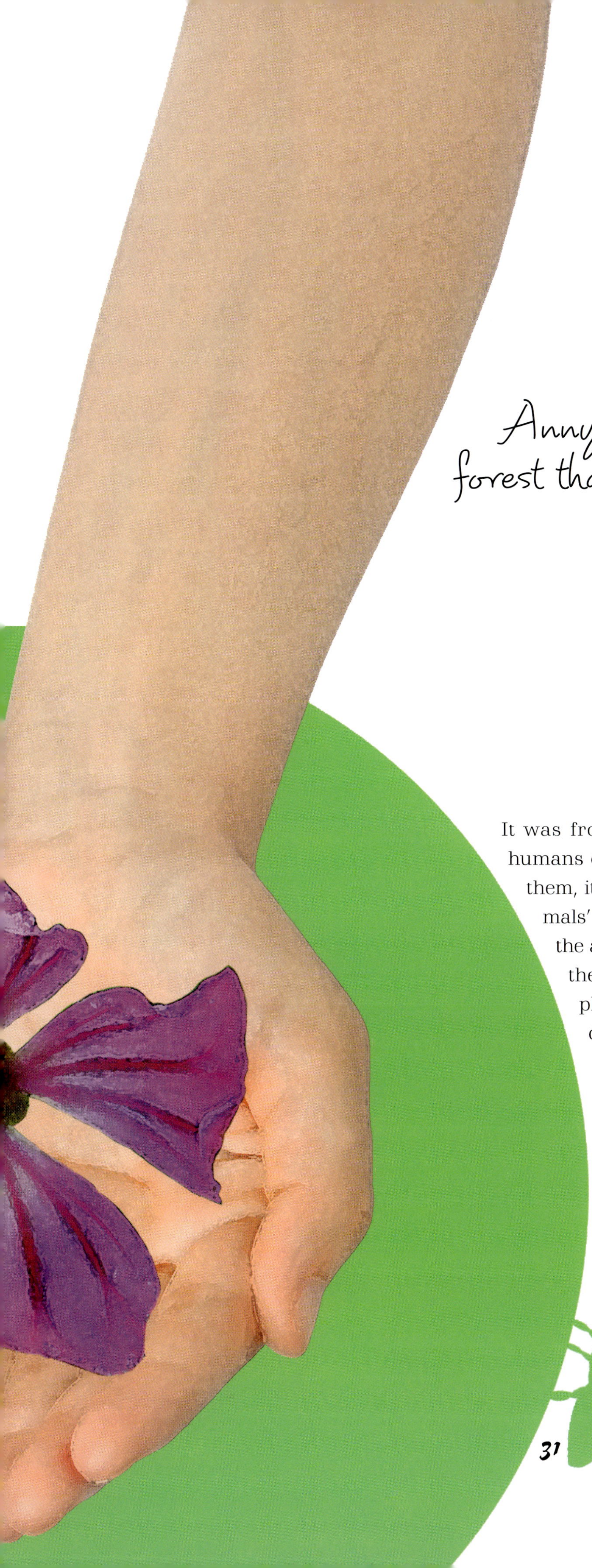

Anny, talk to us about the forest that nourishes and heals...

It was from their observations of animals that humans drew the conclusion: "If that is good for them, it might be good for us too." Because animals' senses were much more acute, people let the animals be their guide as to which plants they could use. That influenced how some plants got their names: dogtooth, which dogs eat in spring to purge themselves of liver stones; and catnip, which tomcats roll in and eat to stimulate their virility.

Regardless of whether the names of plants are derived from them or not, all animals heal themselves naturally. Cows treat themselves with red clover and couch grass, graminaceous plants whose vegetable proteins complete those found in legumes and help them stay healthy; birds are attracted by small fruits that they need for their survival; bees choose various therapeutic nectars to nourish their hives, their larvae, and the queens... This is only one example of the importance of biodiversity, and how its impoverishment threatens many different species. It might even explain the present collapse of bee colonies around the world.

OUR RECENT HISTORY After they were delivered from scurvy by the Amerindians, the colonists soon rearranged the natural order of things. On the plains, they introduced many European plant species. In fact, 80% of the plants that grow there today—dog grass, dandelion, plantain, mallow, red clover—are naturalized plants. It was our ancestors—colonists, apothecaries, and Jesuits—who, whether intentionally or not, imported them. The first recorded colonist on Île d'Orléans, Louis Hébert, was an apothecary and herbalist! Samuel de Champlain himself decreed that each family had to bring eight kinds of plants, seeds, or roots, which could heal or nourish, with them, to ensure the survival of the colony. In general, a lot more was known about plants at the time. They were accessible and free; they constituted the primary medicine of the people.

But the wealth of local resources became quickly known, and that is how many species, like ginseng, almost disappeared. Now, they can only be found in old-growth forests.

Today we are at a turning point: people realize that certain plants, animals, insects, and amphibians exist only in mature forests and nowhere else. Indeed, the only truly indigenous plants we have—and this is equally true of trees and even animal species—are those we find in our mature forests and wetlands.

TREES A mature forest is one that has reached the peak of its growth and development. The trees that have survived are the stronger individuals. At the end of their lives, they fall and lay on the ground, making way for offspring. But even when it is dead, a tree is never useless because it provides shelter and sustenance for various forms of life.

When an oak tree dies A mature oak, for example, is home to about 50 living species—squirrels and other rodents, birds, larvae, mushrooms, and plants that can live only where there is humus produced by oak leaves, as well as gall, which itself is medicinal, and, of course, bark, which has many uses. The oak tree is the quintessential symbol of a mature deciduous forest. It is the tree of the Druids.

A wild forest contains as many indigenous plants as bushes, berries, and non-woody plants that have both medicinal and nutritional properties. Everything has its use. Like *Coriolus versicolor*, a mushroom that grows on old stumps and looks a bit like a tongue. When chewed gently—it tastes of wood and the forest—it can regenerate tissue. It is used to heal maladies of the throat or lungs, as well as to treat abnormal tissue growth. It is a good example of the Doctrine of Signatures, a herbalist philosophy that states that

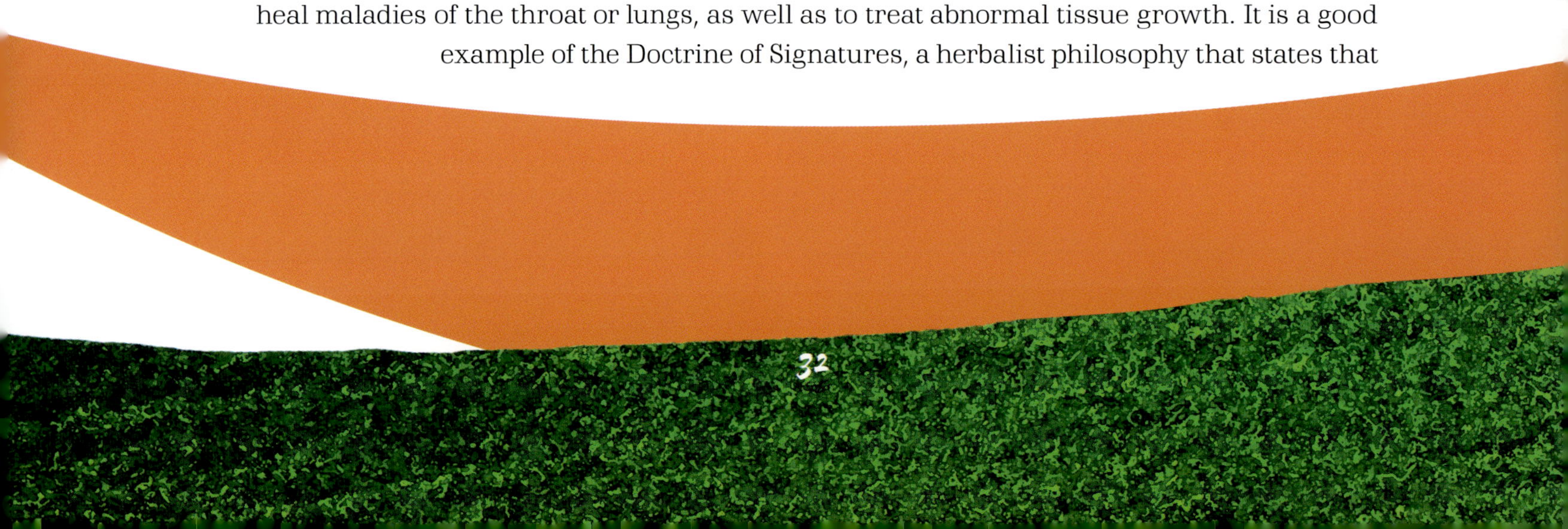

herbs that resemble various parts of the body can be used to treat ailments of that part of the body.

Beneficial hemlock We can also talk about organoleptic analysis—a way of testing using the senses: touch, taste, and smell. Because you have already eaten it, you know that Vitamin C tastes acidic. The young sprouts of the hemlock tree, our national conifer, are extremely rich in Vitamin C, and they can be eaten. Tender, fragrant, and slightly tart, they do wonders for the lungs, throat, and intestines. As well, their essential oil contains a number of anti-inflammatory terpenes. Their tannins are used to tan animal hides.

The fir tree, the king of the forest The fir tree is utterly magical! It is no coincidence that the fir symbolizes eternity, our link with the cosmos, and, of course, Christmas. This tree is very widespread in Canada, but did you know that we export balsam fir essence all over the world, or that it was adopted as the emblem of New Brunswick?

From the gum of fir trees comes one of the best cough syrups ever made. In case of warts, boils, or other cutaneous infections, simply gather the sap with a small knife, put it on a bandage, and cover the wound with it. Each time you change the dressing, clean the wound with oil—the solvent for fir gum. A traditional

enema to evacuate parasites was made from fir gum mixed with egg yolk. Because of its antifungal properties, a decoction can be used as a vaginal douche to combat fungi. It can also be used for foot baths and a host of other things.

There is no doubt that, thanks to the Huron shaman Domagaya, it is these same Vitamin C-rich fir trees that saved the first colonists from scurvy. Or was it cedar? Or spruce? Historians still have not agreed.

Whichever one it was, these three trees are currently under investigation at Université Laval for their anti-tumour properties.

FLOWERS Healing flowers are numerous and abundant. They sprout in the bends of country roads, in glades in spring, and in fields bathed in sunlight. Some are used in tinctures, others are comestible. All are fascinating.

Mallow, for example, is a good mouthwash and can be used as an anti-wrinkle cosmetic, while St. John's Wort makes a calming herbal tea and, as an oil, soothes a sunburn. Ever-popular mint is used in syrups, lozenges, toothpaste, shampoo, and cleaning products.

Bee balm Bee balm is most often found on the edge of clearings, on hilly terrain. Native to central North America, bee balm is a spectacular flower. The perfume of its hermaphrodite flowers is honeyed and mellow, a mixture of lemon and camphor, and its flavour is sweet and mentholated. The Oswego Amerindians who lived south of Lake Ontario used it for centuries to treat food poisoning and intestinal parasites. The Cherokee made a moisturizing, anti-flea ointment out of bee balm heads stewed in bear grease. Analgesic, digestive, antipyretic, antibiotic, anti-inflammatory, parasiticidal... its benefits are endless.

Skullcap Skullcap is found in abundance on the banks of rivers and ponds. In many Amerindian tribes, skullcap was an ingredient in the mixture of lunar herbs given to women to help calm and purify them during their isolation in the menstrual hut.

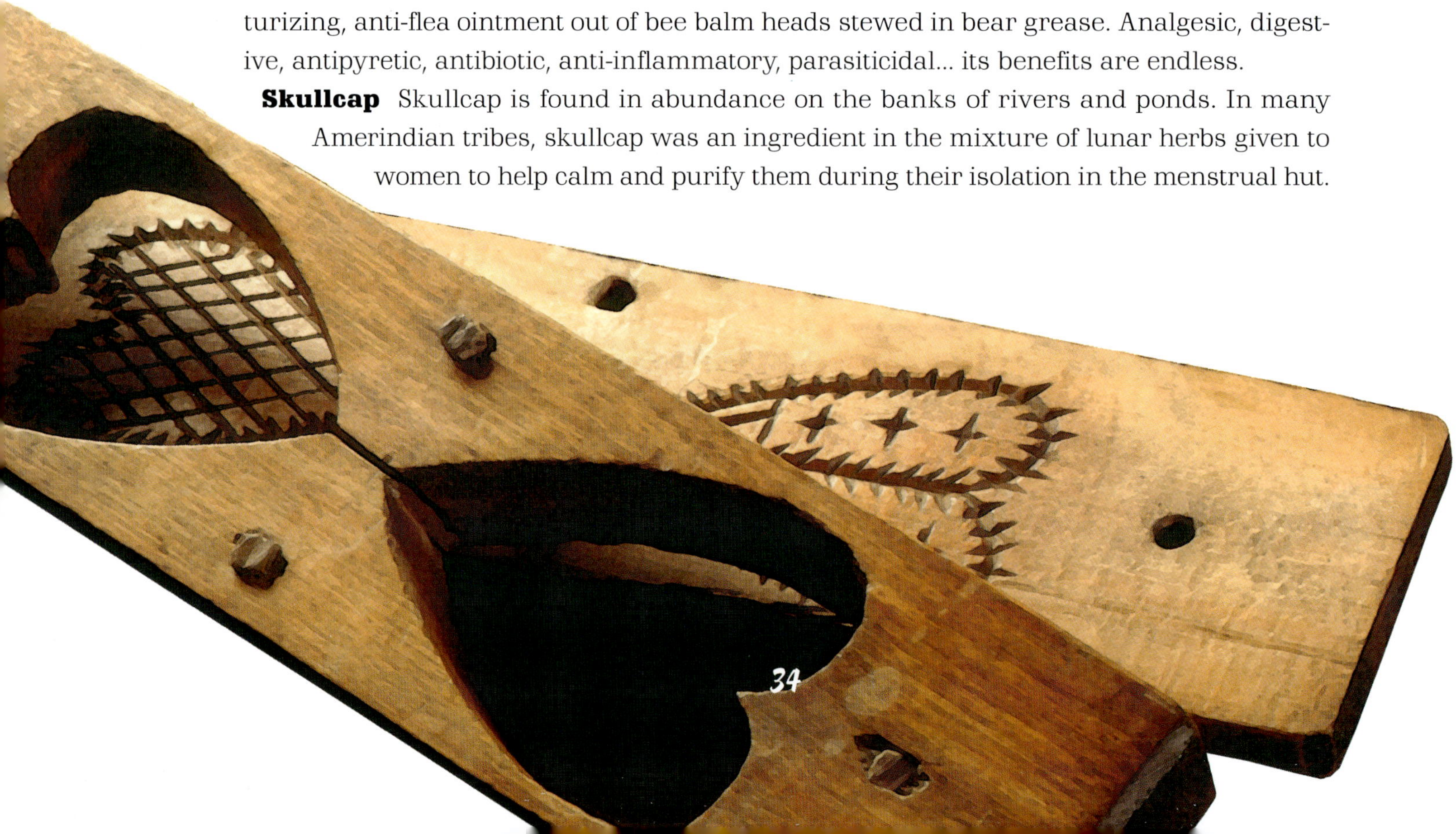

It was also used when sexual taboos were broken, and by the Cherokee against swamp fever. It is still used today to soothe strained nerves and to help break the dependency on nicotine and other alkaloids.

Elderberry A deciduous shrub that grows very quickly, black elderberry thrives on shorelines and on the edges of brooks, although it will grow happily anywhere, with the exception of large coniferous forests. The elderberry bush has always been considered magical. Infusions made from its flowers combat fever and stimulate the immune system.

Did you know that about 80% of all the medical ingredients used in the world—*materia medica*—is obtained from plants. The next time you are in the forest, open your eyes, breathe deeply, and take a closer look at the wonders of the forest, one tree, one plant at a time...

Leisure in the forest

Sylvie, talk to us about leisure in the forest...

THE DEVELOPED FOREST Besides hunters, adventurers, and extreme sports enthusiasts—those ready to venture into the forest with a compass, a topographic map, and the minimum equipment required not to get lost and to be completely autonomous—few people would be able to access the forest had it not been developed.

In Canada, 45% of the land is forests. National parks, which were created to preserve nature while offering large leisure sites, account for 2.2% of the area of the country. Add to that provincial parks—such as those run by SÉPAQ (Société des établissements de plein air du Québec)—and private forests where people hike in summer and cross-country ski in winter.

Although 70% of forests are public, we must understand that large areas are not immediately accessible. When we talk about recreation—as opposed to tourism—proximity is an important factor. That is why parks were developed around major population centres like Toronto and Montreal. For some cities, such as Regina, the creation of urban forests, for sports and outdoor activities, walking, enjoying the scenery, or simply returning to nature, has been a key element in their recreation strategy. They are investing in the psychological well-being these environments can provide.

In a highly dense urban setting, where vegetation is very limited, the forest has special significance for people. Gatineau Park, for example, was born in the city. Nature is abundant there—you can even see bears and other wild animals, which is astonishing, given its immediate proximity to an urban milieu.

Much care is taken when developing these spaces that have dual vocations of leisure and conservation, to make them accessible to everyone, while minimizing their environmental footprint. There is something of interest to admire along even relatively short paths, if only the natural surroundings. When you are in a forest, you also want a larger vision, a sense of opening. If someone wanted, and were able, to walk all day long, they would encounter a variety of landscapes. The trails have been designed this way and each season has been taken into account. Particular care has been taken in considering winter. Where people cross-country ski or snowshoe, everything must be well-planned. Because safety is a primary concern, it is important to make sure people do not get lost, that they can relax while they are in the forest, without having to constantly be on their guard.

Signposting also requires careful, balanced planning. Everything must be clearly marked and indicated, without overdoing it, because this is nature after all! Too many signs would only create needless anxiety for the less seasoned.

The same care that is taken with signage is also devoted to education, to encourage people to stay on the pathways and help them understand that this is to preserve fragile nature. Many activities are offered to raise people's awareness of nature, the diversity of trees and plants, and of birds and other animals. Because the better we know the forest, the more we learn to cherish it, and the more likely we are to respect and protect it.

A HOST OF ACTIVITIES...

In Quebec, almost one person in two practices at least one activity that is related to the forest, or to nature in the larger sense. In the Yukon, we are talking about the entire population! The number of facilities is increasing everywhere: trails for hiking, snowshoeing, horseback riding, cross-country skiing... This increase in the development and management of recreational activities in the forest reflects a growing practice. As an example, the Fédération québécoise de la marche recorded an increase of more than 30% in the number of kilometres of trails that were created between 2002 and 2007, and a surge in their own membership by 47% between 2001 and 2005.

People have not only different abilities but also a variety of desires, including running, biking, walking in a quiet environment where they can listen to the birds without hearing the noise of a motor... Rhythm and intensity vary, depending on what they want. People can enjoy the forest with their family, with friends, or alone, and take part in a sporting activity, or just commune with nature by themselves. The possibilities are just short of endless.

Thrill seekers hurtling across the rugged terrain on mountain bikes or shooting down raging rivers in kayaks share the forest with others out enjoying a leisurely stroll at their own pace, relaxed, camera in hand or sketchbook under arm, admiring the beautiful fall colours or the animals... It is there for everyone. Someone who simply wants to walk in the forest without any special equipment can derive as much pleasure as the accomplished athlete in search of new adventure. Everyone can find their share of happiness in the forest.

...AND EXPERIENCES The forest offers people not just one, but a whole range of outdoor experiences. Our experiences vary depending on what we are looking for.

The sensory experience, for example, is vastly different from what city dwellers are accustomed to. In the forest, you do not control the space. You may run into other people or not. You have to pay much more attention to what is going on around you. You do not speak loudly, especially if you want to see and observe animals. You change in some way, and become more sensitive to smells and noises. People who walk in the forest experience sensations they had not been expecting. Others, sensing the freedom the forest affords, even touch the trees, in a desire to become one with the forest.

When you fish or hunt, keen observation and hearing are crucial. You are on the lookout. You have to be attentive to every sound, every movement. Your senses are sharpened in the forest. If you are kayaking on a lake, for instance, and a loon swims by, suddenly, everything is intensely focused. Longing to see it again, you paddle after it. It is as if you were on a hunt: your vision is keener, and you pay attention to the smallest sound. This is far removed from the deafening bustle of urban life!

RELAXATION, HEALTH, AND HAPPINESS Walking in the forest is one of the simplest forms of leisure activities that brings a great deal of satisfaction. It encourages people to relax and refocus. It promotes not only a return to nature but also to one's inner self and origins.

A long time ago, philosophers would take long walks to help them think. Writers have touted the benefits of walking in nature to liberate and stimulate ideas. It is as though, when the body is in motion in a conducive space, thoughts arise spontaneously in bursts of inspiration, and everything becomes clear.

Walking at a moderate pace is of course in itself an activity that promotes good health. We often hear that 30 minutes of walking every day is enough. But out in nature, people tend to walk for longer because the surroundings are so pleasant, and this encourages us to extend our activity.

When you walk in the forest, you are not really aware of the distance you cover, nor of the time that passes. You walk at a different speed in the forest than in the city, where noise and traffic incite you to move faster. The city imposes its rhythm on you, whereas, in the forest, you set the pace: slowly or quickly, it is up to you. Along the way, you can pause in any number of beautiful spots. This brings a feeling of calm and gives you the chance to collect your thoughts.

What about the myriad views? On a lake, for instance, you can look at the forest from the water, admire the diversity of tree species,

and contemplate the reflection of the sky and forest on the water. The experience is quite magical. It is really worthwhile to vary your leisure experiences in the forest: appreciate it not only from within the heart of the forest, where you can tap into its pulse and commune with it, but also from afar, in a kayak or a canoe, from the water, where you will be rewarded with spectacular and breathtaking views. You will be filled with wonder, and you will say to yourself: "This is magnificent! I will never get tired of this."

The forest that celebrates

Marie-Andrée, talk to us about imagination and creation...

TIME AND THE DISTORTION OF TIME

When I walk into a forest, I feel as if I am magically transported from the hustle and bustle of modern life, with all its problems, to a world that is timeless, totally different, superimposed, as it were, on daily reality. I enter into silence. I am drawn in.

The images I see are probably the same as they were 300 or 400 years ago. The trees are like elder sages, witnesses of time, protected in their domain. That makes the forest feel like a safe place, conducive to meditation.

And yet, I always feel like I am arriving on some sort of a set, without knowing exactly what to expect. It is as though Mother Nature had prepared a special show for me, one that changes every time, because the lighting and the costumes are never the same. It is a spectacle that is at once spontaneous and grandiose.

Through the changing seasons, we savour the rhythm of life. In the forest, everything is both slow and fast. Slow, because with so many things to stimulate the senses, each moment can seem like an eternity, and fast because things move more quickly than you could imagine possible. At the end of summer, for example, you can stand dumbfounded beneath a tree that was covered with green leaves just one week before; it has suddenly begun to turn red, a harbinger of the change that will not be long in coming. Over the course of time and through all the years of transformation, the tree remains. The forest is there to remind us of the passage of the seasons. But we tend to forget about it, because we live in a world where too many things distract us, a world in which we consider ourselves immortal. We have to remember that life is a cycle, and we should relish every moment.

ELEVATION The tree is more or less our symbol. Planted on earth, a tree strives to rise up towards the light. It seeks a passage, as if trying to understand. It seeks the most advantageous position. Sometimes, I feel a like I am pursuing the same things, that I am on the same quest.

We can also attribute motive to trees. If I go into the forest in need of comfort, for example, the trees seem to come together like a gathering of old soothsayers to offer me solace. When I lean against a tree, I immediately feel its soothing energy. If, another day, I have had some good news and feel like celebrating, all the leaves flutter around me, and I imagine a host of little hands applauding me. It does not matter what I feel, the trees always accompany me, no matter what state I am in, no matter what my needs are, no matter what I am looking for.

In the forest, in the face of such a colossal mass, you discover a humbler, more intuitive side of yourself. Amongst one another, people often try to take control. But when we are alone, deep in the forest, we feel small, touched by such magnificence.

IMAGINATION What nourishes imagination?

Take a walk in the forest when you feel tired, distracted, or overwhelmed... you will suddenly feel calm again. Harmony. Cohesion. Beauty. You realize your days do not leave much room for imagination. You do not suddenly stop in the middle of everything and say: "Hold on, I am just going to enjoy a moment of imagination!" That is not likely to happen, since in this world—outside the forest—everything has to be organized, everything has to be in working order and has to move quickly, while imagination demands time, space, and calm. Imagination is not a recipe or a painting-by-numbers. Taking the time to imagine means first connecting with yourself, with your emotions. And, too often today, that is a luxury, isn't it?

Imagination is the spark of an idea. The world of imagination is a private world. Ultimately, there is nothing more personal than the imaginary world, because it is based on our representation of things. You cannot be more in harmony with yourself than when you are living an imaginary moment, where all available information and everything you can see, feel, and hear passes through the prism that is you, expressing itself deep within you.

I am sure that the ocean can also help us access these depths of imagination. Deep underwater, the imagination cannot help but be stirred. But that environment is, to me, less reassuring, because it is physically foreign to me. The forest is where I feel safest, most protected, as much from the whirlwind of modern life as from the rain and heat. It offers me wonderful moments of introspection.

CREATION What stimulates me in my work is the changing colours of the seasons and the forest, which trigger the artistic process and provide me with visual material. In my studio, in my paintings, I convey the emotion of my subject through my colour choices.

A fall forest, where all the leaves have turned brilliant red, for example, will knock you out with its powerful symbolism. Today, red means something different than in the past, due, in large part, to the constant barrage of images of war served up by the media. In the past, red meant love! Any elderly person will tell you that! Now, it brings to mind armed conflict and blood, and it evokes violence and injury. If you find yourself in a crimson forest, then, where colour is suddenly embodied in matter, and where the play of light imbues it with a deeper meaning within that peaceful environment, you are reminded that life is a cycle… Appreciate it while it is there, because soon all this red will soon disappear and return to the earth, and the elements of creation will renew themselves once more.

Another entrancing experience is strolling through the forest after the first snowfall. It looks as if a giant hand sprinkled a dusting of flour everywhere. And when the first snow falls, the trees still have leaves, so the snow sticks to them in places, but not evenly. It is utterly beautiful, truly magical.

And what about light as a source of inspiration? At dusk, the sky changes from purplish blue to pink and deep mauve before turning indigo blue. But it does not last long because the light fades quickly, especially in winter. In the morning, the slanted rays of light always fill me with wonder. They are almost horizontal when the sun rises. It creates the impression of little crystals of light vibrating on the snowflakes in a kind of ballet. Fairylike!

The beauty of nature and its astonishing harmony is a work of unequalled perfection, but, too often, we forget to admire it. We easily lose sight of the fact that there is something extraordinary, something greater than ourselves right at our feet and in front of our eyes.

The painter Marc-Aurèle Fortin was fascinated by trees, which occupy a central place in his work. What is remarkable in his work is that, for him, there was never any disconnect between content and form. He painted a hole in a tree like a stain in the sky, because everything is part of the whole, the same universe, all related. His paintings are very powerful, full of emotion. This is the emotion that every artist tries to capture and convey.

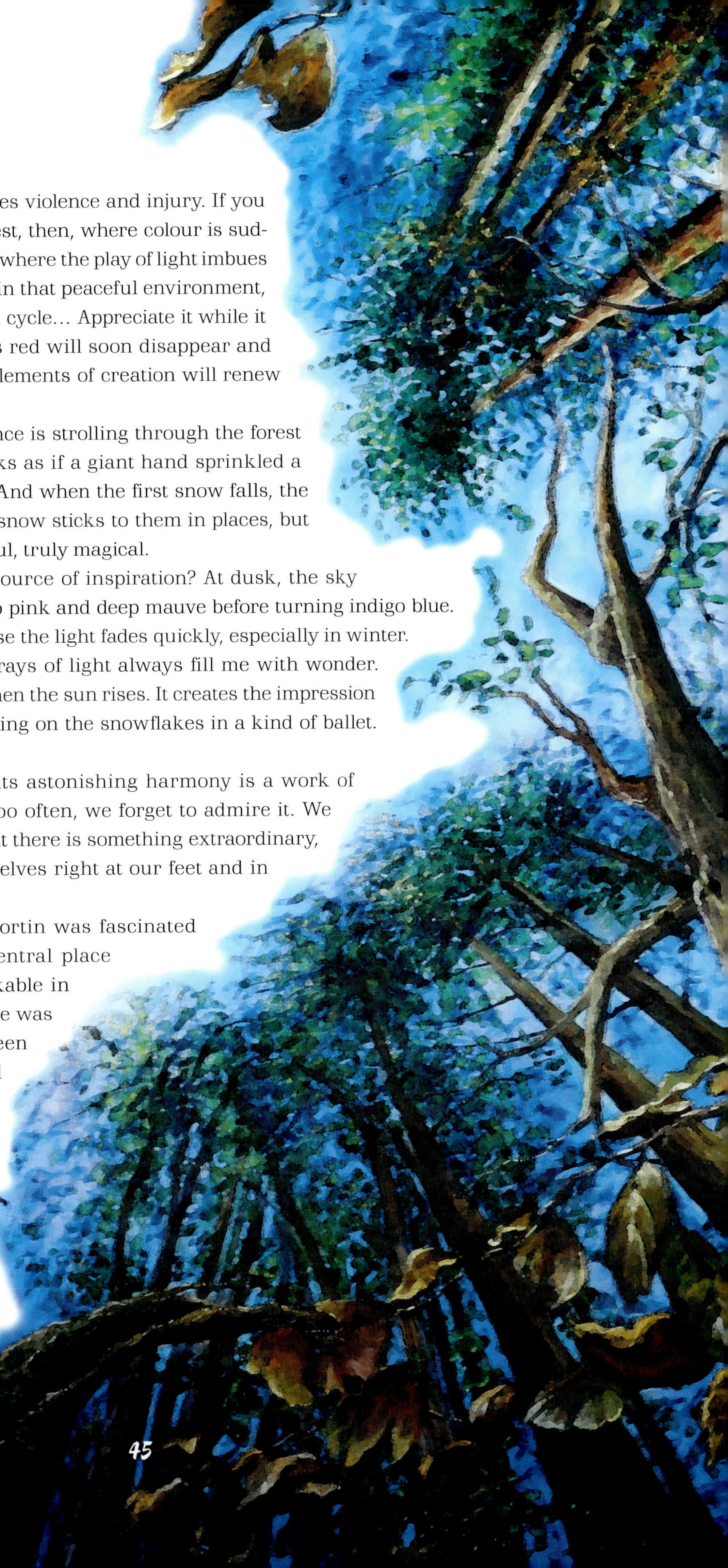

The appeal of the forest

Julianna, talk to us about ecotourism and the forest...

THE LAURENTIAN FOREST My background is half Australian, half European. I have just spent seven years in Canada, working as an expert for Parks Canada. I love the outdoors, and I can say that the best nature experiences I have had have been in the forest.

In the course of my work, I got to know and appreciate the Canadian forests better, particularly the Laurentian forest in Quebec, primarily as a natural resource, but also as, we sometimes forget, a strong cultural one. If we look at the tourism offerings in Quebec and elsewhere in Canada, we can see that, as far as nature tourism is concerned, the forest tops the list. It is easy to understand why, because the entire country is covered with forest! Every built or cultural environment is located in, or close to, a wooded area. This makes the forest not only a natural resource, but a very important economic and cultural one as well.

But beyond that is its sheer magnificence. In terms of landscape, the forest is a veritable wonderland, offering people countless opportunities to experience nature. Depending on the time of year, that experience can be absolutely heart-stopping, such as in the fall when the colours are on display. I used to work for Parks Canada, in the La Mauricie National Park. For a number of years, we conducted a series of evaluations of the visitor experience there. The goal was to improve and enrich the tourist experience in this protected area. That was the first

time I had been in a protected park, a Laurentian forest, in the fall. The colours were at their peak, and the leaves were... indescribable!

The experience was so extraordinary, so inspirational and like nothing I had ever had before, that I will never forget it!

AN OUTSIDE VIEW As an ecotourism specialist, I am especially interested in everything that has to do with the environment, tourism, protected areas, and the sustainable development of the land. I do not live in Canada any more, but the first thing I think about when I remember the country is the forest. In my mind's eye, I see the woods, the animals, the lakes, the typically Canadian rustic cabins... all those things we consider commonplace when we have them at our doorstep.

Since I returned to Europe, the work I have been doing offers me vastly different experiences. It is certainly beautiful here, but it is not like Canada, where great swathes of land are covered by natural forest, which, in most cases, have been scarcely or never touched by human hands.

This is a wealth that is not always fully appreciated, because we tend to take it for granted. When you live there and work in the tourism industry, it is easy to forget that the huge, sweeping expanses we find in Canada are a tremendously rich resource with enormous potential.

I have travelled the world, and I have visited many countries and regions and, in my observation, Canada's ecotourism spaces are very well maintained. Many areas still have virgin forest, and that is quite exceptional.

In Africa and South America, the forests are beautiful too, but they have been more disturbed by human activity. In Europe, I walk in the forest every weekend, but I can never experience complete silence. In Canada, in Quebec, you can still find quiet spots in the forest or by the side of a lake, where you hear nothing but the sounds of nature. This has value that cannot be quantified. Places like this are increasingly rare on the planet!

TOURIST OPTIONS Ecotourism represents a substantial part of the tourist economy. We have stopped counting the number of businesses that offer outdoor tourist activities—kayaking, canoeing, hiking, camping—operations that are all essentially related to nature and the forest.

Incidentally, you might think that hunting and fishing are purely local activities. I analyzed the various ecotourism offerings in Quebec, including hunting and fishing businesses as well as outfitting operations in general. Among my findings were that increasing numbers of people are taking part in these activities, and many are using the services of an outfitter. The market seems to be equally divided among Europeans—who are more interested in the cabin-in-the-forest experience, a little hunting, and group excursions—and the Americans, who have long been fervent hunters and fishermen. And remember that, in the past, a number of wealthy Americans established hunting clubs in northeastern Canada and the United States...

There is great potential in ecotourism, in the areas of hunting and fishing, and because it attracts foreign visitors to Canada and gives them the opportunity to experience the forest all year long. Whether they come from Chile, Mexico, or Switzerland, they are all looking for a forest experience that brings them into contact with true wilderness, breathtaking landscapes, and vast expanses of nature.

In winter, we are seeing the re-emergence of new trends from old: snowshoeing and cross-country skiing. Snowshoeing, especially, has become very popular almost everywhere in the United States, Canada, and Europe, partly because it does not have an impact on the natural environment, but also because it is a healthy activity!

FROM SYRUP TO THE TREE People who make their living from the forest are benefitting from this trend toward outdoor activities. Visitors come to see animals in their natural habitat, or, in the spring when the snow begins to melt, to enjoy an outing to a sugar shack, where maple syrup is produced. This local resource is unique in the world, and maple sugar aficionados come from as far away as Japan!

It goes without saying that the forest is full of activities in the summer, while the fall is a time for admiring the colours. The result is a cyclical offering of activities of all

kinds that provide visitors with a diversity of ecological experiences in the forest and brings them closer to nature.

As far as the protection of natural areas is concerned, this summer, the European media reported that Canada was going to increase its protected areas, to preserve its natural environment, notably the boreal forest. This is indeed excellent news that more people need to know about. It is a huge tourist offering, one that must be protected to ensure that it will always be there for future generations to enjoy.

Canada offers an exceptional window onto the forest.

Wood from the forest

Hervé, talk to us about the wood that comes from the forest...

THE PLEASURE OF WOOD What is it about seeing and touching wood that comforts and soothes us? Why is it that when we see a piece of wood, we are drawn to touch it?

When I was a child in Saint-Jean-Port-Joli, we would get our knives sharpened by sculptors. After years of whittling the same quaint little figures of old men and women perched in front of fireplaces, these artisans continued to experience a renewed pleasure each time they touched a piece of wood.

And it is not just pleasure that is renewable. Wood is too! The forest is home to birds, deer, and moose, and even though we exploit the forest for its wood to make products to satisfy people's needs or desires, the animals can continue living there.

This interaction between the forest, wood, and man is fascinating. My grandparents were forest contractors and maple syrup producers. They took care of the forest. During the war, my grandfather cultivated large maple groves to produce sugar, because sugar was very hard to come by at the time. The relationship between the forest, wood, and human beings has something of visceral quality to it.

What can we do with wood? What sensations can it create? The answer is that we cannot begin to imagine everything that can be done with wood, nor all the feelings it can arouse. Think of all the fascinating ways we use wood, for necessity as well as pleasure. In both cases, we are now experiencing a veritable boom with a multiplicity of applications. Let me explain how the forest is a source of many other resources besides wood.

AN IMPORTANT WATER REGULATOR At first, the energy used to cut and transform wood was provided by waterfalls—hydraulic power. Then we moved on to electric motors, and people began building dams and producing electricity destined for transformative activities and a thousand-and-one other uses.

But the relationship between water and the forest goes well beyond energy considerations. In fact, the forest is our most important water regulator. In addition to being a producer of water, a well-managed forest purifies the water as well, preventing it from flowing directly into watercourses. In other words, it replenishes the ground water.

This highlights one of the problems that exists in agricultural zones. Historically—and one can easily appreciate why—settlers first claimed the richest lands in the St. Lawrence plains, and proceeded to drain them. The result is that now, the day after a rainfall, the water has already made its way into the river along with sediment, nutrients, fertilizers, and all. This is why we are currently working on trying to find ways to encourage the water to percolate in the cultivated soils rather than letting it escape into the ground water.

The forest fulfills that function naturally. It acts like a sponge from the moment water touches the ground. And when the water passes through soil, it is cleaned. In India, in the Rajasthan region, I participated in projects to rehabilitate river basins in areas that had been largely stripped of trees. The result: our work allowed the water table to rise by 20 meters over five years. By planting trees and erecting intermediary dams, we built jetties that created artificial lakes in the streams and small rivers, which allowed the water to percolate in the soil.

In Canada, we could produce drinking water. It is a rich resource that has not yet been fully exploited. We would be able do this through the judicious use of the forest.

A FACTOR FOR SUSTAINABLE DEVELOPMENT When the Europeans arrived in North America, the first thing they did was to make tools and build shelters for themselves. To do that, the craftsmen had to make use of available materials, notably wood.

Houses from that period still exist today, and remain in very good condition. If we have been able to preserve this heritage, it is because wood, in addition to being functional, is also durable.

Development in Quebec began with the creation of facilities to process raw wood. People then went on to build houses, barns, churches, carriages, and agricultural implements. Back then, those were the needs. Thousands and thousands of sawmills sprung up—on seigneurial lands to begin with—and they eventually developed into small-scale businesses, which, along with forges and flourmills, could be considered the first small- and medium-sized enterprises. Communities established themselves around these industries, which provided both the essential necessities as well as the conveniences of life.

When towns, such as Montreal and Quebec City, started to grow, more wood was needed to build them. Highly effective construction techniques were developed, with supporting structures that were light, allowing one-, two-, three-, and four-storey buildings to be constructed completely out of wood. They were so well built that they still stand today.

And so, an industry of necessity became a local industry, then an exporting industry. Over the last few years, 80% of our wood has been sold to our American neighbours to support their construction market. Why? Because we are able to produce lumber from small diameter trees from the boreal forest. Research has led to processing methods that can produce 2x3 and 2x4 lumber, which is the basis of building support structures.

TODAY... We have added value to our forests. We now understand that the forest has an influence on climate and on carbon levels, that it is a place for recreation and leisure, and that it plays an important role in the conservation and quality of water. In the future, we should seek a balance between forest exploitation, conservation, and recreation.

In the province of Quebec, due to moderate exploitation, the forest is in better shape than it has ever been. Our forests are rejuvenated, and, therefore, more productive, more biodiverse, and more interesting. The boreal forest and mixed forests have a phenomenal ability to regenerate. In fact, the problem here is not a lack, but rather an overabundance of trees.

When a forest is cut down, it triggers extremely dense regeneration. Trees compete with each other for a long time, until the dominant ones gain the upper hand. There are approximately between 35,000 to 40,000 seedlings in a hectare, where only 200 mature trees are needed. This is an enormous advantage over the rest of the planet: we have a vast and diverse forested area, with a large variety of species that contribute to keeping the soil rich.

Because of our cold climate, among other factors, the ground will usually be covered with shrubs less than five years after a cut. In the tropics, the forests do not accumulate humus. They draw a lot of resources from the soil. Without a supply of humus, the soil rapidly becomes impoverished, exhausted, and forms what is known as a hardpan, where nothing, or almost nothing, grows. Here in Canada, when the forest is harvested, seeds in the humus, the black earth on the ground, are activated. Thus, biodiversity is assured.

For us, the challenge has never been development, but rather the use of wood. When we find applications for different diameters and different species of trees, instead of working against nature, we work with it. If we exploit the forest on that basis, we will not have to continually keep sacrificing land, or trying to grow species that refuse to take root naturally. We therefore have to find uses for wood that let us take advantage of all species, all qualities, and all parts of the tree, and harvest wood with the goal of producing a sustained yield, which will allow the forest to regenerate naturally.

...AND TOMORROW We understand that the future of the industry will rely, to a great extent, on the value of raw lumber used for construction materials, among other things. Structural products and a new generation of appearance products will be integrated into two different business models.

The first model is prefabrication. In the near future, 90% of all family homes will be prefabricated, because of the ease and rapidity of construction of this method. Thanks to new technologies, it is almost possible to inhabit a new house before it is built, because we can use computer software to choose the materials we want in a house, as well as the furnishings and decorative elements. In this way, we are adding an element of engineering, and therefore value, to the project.

The second model concerns the non-residential market of office and commercial buildings. The forest industry had backed away from this market, because, until now, the demand for commodity products has been so high. The business model for these basic products was certainly more appealing, but it did not create the most value, nor did it generate the most knowledge-based jobs. But today we have the example of the Fondaction building in Quebec City that features a hybrid structure surrounded by a wood-frame structure—a first in North America. Since then, another building with a light, reinforced framework—over two storeys high—was constructed in Vancouver.

We are currently developing products that will allow us to put together buildings a little like assembling Lego® blocks, as rapidly and effectively as using steel and concrete. If they last, if they promote energy conservation and add value and knowledge while providing the warmth and pleasure inherent in wood... how wonderful that would be!

We are working on every aspect: durability, mechanical and seismic resistance, tolerance to vibrations, fireproofing, soundproofing... We are entering a new phase of history. Canada

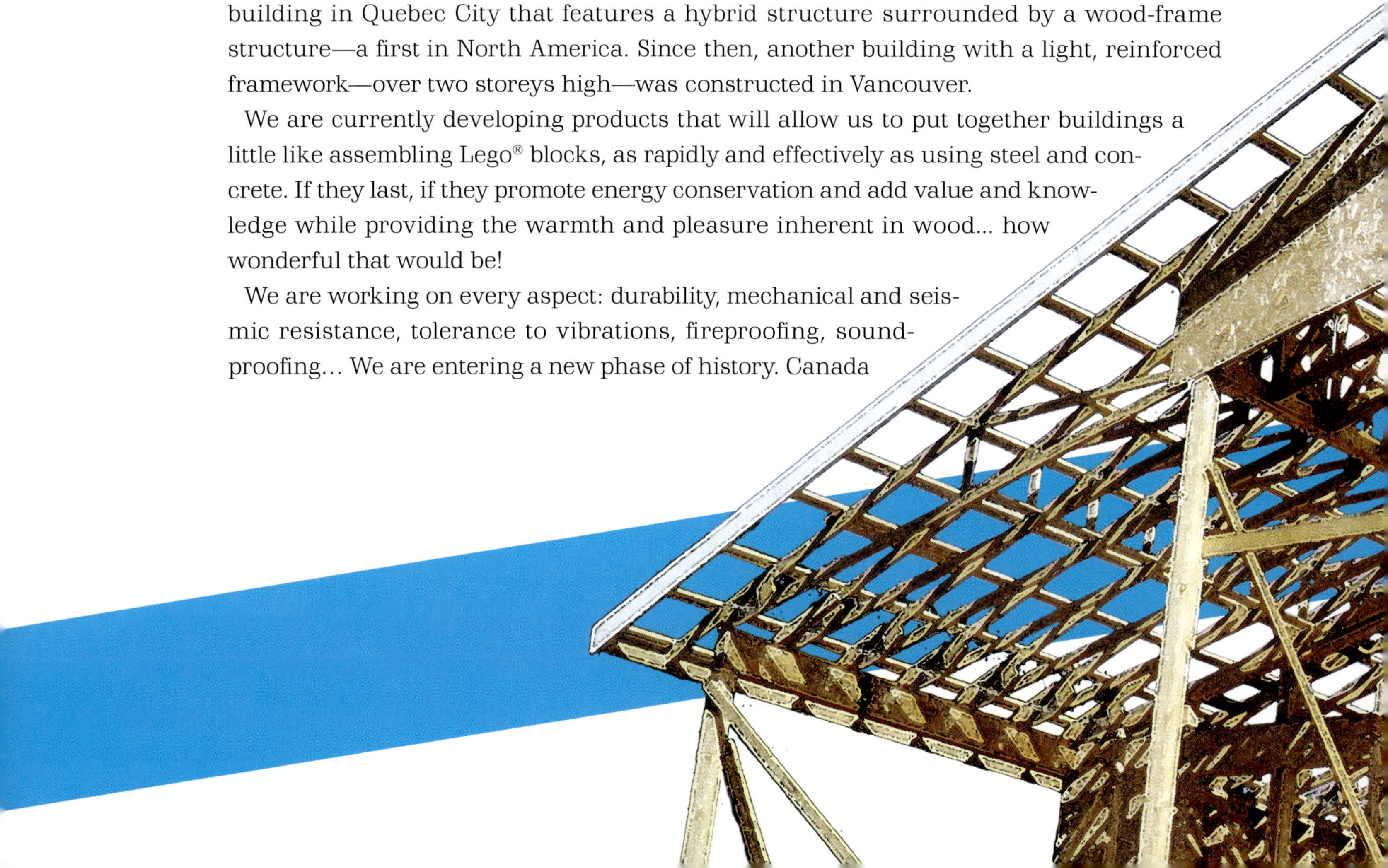

is a leader in this field, through its highly active research and emerging technologies, such as nanocellulose, to increase the durability of wood and provide it with other interesting properties.

A GAUGE OF REGIONAL PROSPERITY For 25 years, the wood and pulp and paper industries have been working in synergy, because, with the procedures that have been put in place, we have been able to refine wood shavings. Historically, paper was made from ground shavings. When that changed to milled shavings, raw materials were processed in sawmills before being used by the pulp and paper factories. The two industries thus became closely linked.

The forest industry has always been, and will remain, deeply rooted in the outlying regions for a very simple reason: the transportation of wood is an important factor, and it is always in these regions that wood is easiest to transform. Because we give our products increasingly higher added value, this will likely continue to be the situation in the future.

The wood industry employs about a quarter of a million people in Quebec and represents the second or third largest export industry. The health of the economy depends on our ability to develop the entire territory, and the forest industry accounts for much of it. Did you know, for instance, that 90% of the roads in Quebec were built by the forest industry? In addition to providing access to regions, the forest industry has become the backbone that supports many communities. It also generates value, mobilizing other industries like transportation or banking. The bottom line is that 250 Quebec towns and villages depend on the forestry industry, and let us not forget that their head offices are in the larger cities!

Did you know? The first nanocellular plant opened in Windsor, Ontario, just a few months ago.

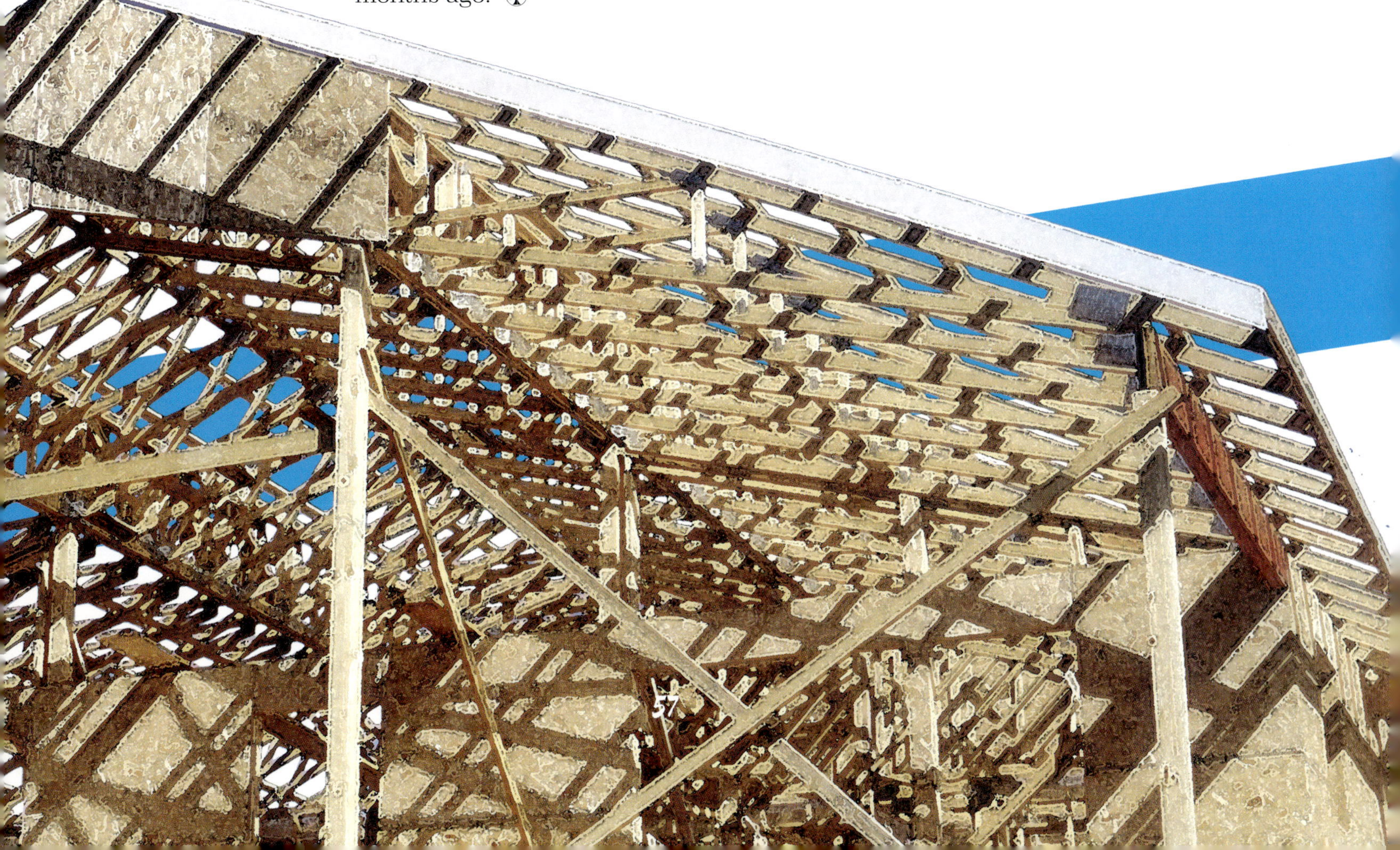

Paper and the forest

Jean, talk to us about paper...

PAPER, PAPER, EVERYWHERE When you talk about paper, you are talking about a vast universe and 600,000 jobs in Canada. Paper is used for printing, writing, packaging... Just think about all the paper products you come into contact with when you have a meal in a fast-food restaurant: the cardboard box, paper on your tray, cover of the cream container, sugar sachet, napkins, bill, bathroom tissue and hand towels in the washrooms... In that short period of time, you can use over a dozen paper products. Paper is an intrinsic part of our lives, an integral part of almost every aspect of our existence.

It is even found in electrical materials, because of its insulating properties. It lends itself naturally to the fabrication of many objects, because it can be recycled. You can take paper, transform it into pulp and make more paper, deink it if it has been printed, and repeat the process up to a dozen times. Not only can paper be reused, but at the end of its useful life, it can be burned to recover the fibre's energy value, thus returning it to nature. Its versatility is its beauty.

The forest captures solar energy, which is retained in paper. A tree absorbs the sun's energy and the carbon in the air, and converts them into a type of sugar that forms chains of cellulose. The tree then assembles these chains of cellulose into wood fibre. This fibre has unique properties that allows it to be used to make wood pulp, paper, and cardboard. The transformation of fibre into paper products allows it to be reused many times, and even to be stored for fifty years.

HOW DOES IT WORK? Paper has a fascinating structure, which you cannot see with the naked eye. To make paper, wood fibre is turned into pulp and dried. The fibres stick to one another—without glue—and form a rigid structure. This rigid quality is necessary in many applications. When you read a newspaper, for example, you appreciate that the pages stay up and do not fall limply. And when you read a book, you want to be able to turn the pages. It is also the structure of the fibres that gives cardboard boxes the rigidity and strength they need to protect and transport products of all kinds.

So, this structure, which is extremely thin—between 20 and 300 microns, really very little—is both rigid (it can be printed) and opaque (you cannot see through it). The Chinese first discovered this property of cellulose, that it forms another structure when it is reduced to pulp and allowed to dry.

Twenty-five years ago, entire trees were used to make paper, but not today. In Canada, the majority of paper mills no longer use roundwood, but the wood chips from sawmills instead. These days, when a tree is felled, its bark is removed and transformed into construction materials. We optimize the squaring, which creates wood chips that are then used to make paper. In this way, all the pulp is produced from the residue of a prior operation. We have learned to maximize the forest's value. And we are continually working on improving the supply chain.

We have also learned to make mechanical wood pulp from wood chips, in order to transform as much of the available woody matter as possible. We make either kraft

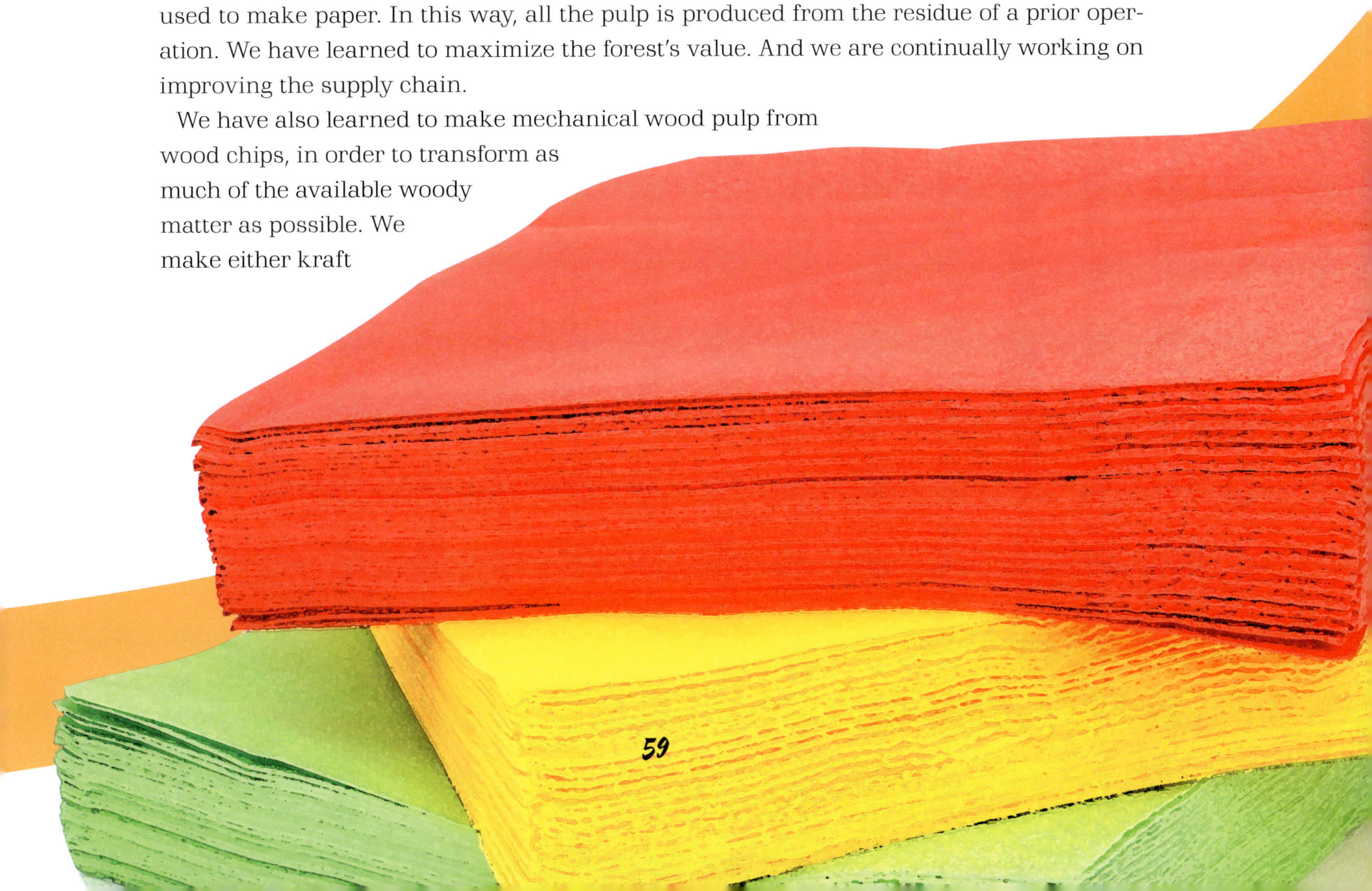

pulp or mechanical pulp. In the case of mechanical pulp, used to produce newsprint among other things, the entire fibre goes into the sheet. The kraft process, which chemically separates the fibres, is used to make fine papers. A little material is lost in this process—principally lignin—but it is burned to generate energy, and many mills sell this form of excess energy to electrical grids.

BIOMATERIALS Paper is slowly evolving towards what we call a biomaterial.

Wood fibre is made of smaller fibrils, which, in turn, are made up of microfibrils. They can be peeled—a little like cheese strings—and the resulting filaments can be used as material. These filaments can be of nanoscopic dimensions, the smallest component of wood being cellulose crystal. We are talking about nanocellulose here. In comparison, one nanometre corresponds to the diameter of a hair divided into 36,000 parts. It is a little like comparing an orange to the earth.

Now, by reassembling crystal cellulose in various ways, we can control the properties of light and, therefore, the colour of the material. In fact, we are in the process of rediscovering the qualities and properties of wood fibre, and, by the same token, finding new ways of using it and rearranging its components into different materials for the construction, aeronautic, and automobile industries, as well as a host of other applications. We are developing wood fibre for more advanced, more high-tech applications, while conserving its natural qualities, and creating materials that are recyclable, reusable, and compostable.

Nanocrystalline cellulose, nanofibrils, cellulose gel... the vocabulary is growing along with the research. And the possibilities are endless. We could, for example, form cellulose gel into plastic sheets or parts, or even toys! That way, 20 years down the road, when the kids have grown up and left home, you could simply put the toys in the garden and watch them turn into compost!

Products like that will soon be appearing on the market.

WHEN PAPER BECOMES INTELLIGENT Using these components, we think we will be able to produce even more sophisticated materials. By combining nanostructures with electronics and conductors, we could, for example, make luminescent threads and weave them into a tapestry that lights up your living room, or "intelligent" packaging that tracks the progress of a food product in a very economical way—origin, transportation conditions, freeze or thaw events, etc. We can also envisage foldable, rollable, recyclable, and compostable display stands.

What is interesting is that many of the basic products will be produced in existing mills. Because trees will still have to be harvested, the wood chips still collected and transformed into fibre, the value chain will therefore be created in existing facilities. This is an ongoing and ever-evolving story.

It is clear that we cannot keep exploiting oil endlessly. We are going to have to replace it with materials that are naturally accessible, recyclable, and reusable. Cellulose happens to be the most available material in nature. The most abundant biomass on earth is cellulose. We have to find a way to process it, to put it to multiple uses, and to reinvent our products using wood fibre, the forest, as their base.

A NATURAL STORY OF EVOLUTION AND PROTECTION Thanks to the forest and to wood fibre, which is at the heart of the entire history of cellulose, paper is evolving. Nature developed a molecule that has assumed myriad

forms throughout its evolution, going back millions of years. To reach the sunlight, trees had to get bigger and grow ever taller—dozens of meters taller. The long and the short of it is that the forests contain huge quantities of raw materials. All we have to do is use them... for other things than campfires!

If we look carefully at a map of global forest coverage, we see that forests do not cover that much of the earth's surface. It is true that in Europe, Russia represents an enormous resource, that there are some wooded areas in Asia, and that in North America, the northeast and the west have abundant forests. But overall, worldwide, not much land is covered by forest. That is why we have a responsibility to take care of ours, and cultivate it as best we can.

Actually, in Canada, we cultivate the forest less than we maintain it. We have chosen to preserve diversity, which is what counts, because we still have a great diversity of wood species: we must maintain the entire ecosystem while we determine what can and cannot be used. We have to harvest the forest in the same way that we harvest wheat to make bread, or strawberries to eat. Farmers see their lands dry out if they overexploit them and do not let them lie fallow from time to time. The same goes for the forest. To ensure sustainability, man and the forest must work together, in tandem.

The energy reserves that exist underground, in the form of petroleum, oil, or gas, come from molecules produced by our forests over millions of years. We are in the process of exhausting those reserves. The only other energy source that remains—and that is renewable—is the forest.

Energy in the forest

Daniel, talk to us about energy and the forest...

MYSTERIOUS ENERGY There is really no universally accepted definition of energy. It is, frankly, a somewhat abstract concept. Energy helps us define how an action—a force or pressure applied to something—can have a certain effect on an object. For example, when you light a fire under a pot of water, the water boils, and steam escapes. You can convert this physical reaction into a unit of measurement by comparing it to a mechanical equivalent. We have agreed to call this measurement energy. To simplify things, we could also illustrate the concept by saying that energy is the product with which we produce all products...

Different sources of energy exist in a natural state. Take a simple tree branch found in the woods. Using the mechanical energy of our bodies, we can put the branch into a stove, set it on fire and use the resulting heat—thermal energy—in various ways. But to do that, we ourselves have to expend energy. Which means that there is always performance to consider, because to produce energy you need to use energy. So energy is a product we have to invest in. We need to remove coal from where it lies underground, and it is costing us more and more to find and extract petroleum.

We have already obtained yields of up to 30 units of energy produced per unit spent. In other words, by providing one unit of energy, we obtain 30. In the case of bituminous sands (tar sands) it is more like 3 to 1, and sometimes even 2 to 1; it takes a lot of energy to extract petroleum. The same can be said of drilling for black gold at the bottom of the oceans: the energy required is considerable, not to mention that the yields are sometimes ridiculously low, especially when thousands of litres an hour are lost in disasters like the one last summer in the Gulf of Mexico.

In short, energy is the mother of all products. It is thanks to the energy a society produces that everything else can be produced: buildings, roads, manufactured products... very little today is produced without it, because everything is so mechanized. Maybe that is why so many people dream of having their own vegetable plot; it corresponds nicely to the amount of energy one human being can invest.

A RETURN TO THE MEANING OF LIFE If there is one fundamental, and very interesting, notion when we talk about energy, it has to be that of horsepower. The power of a modern car, for example, is far more impressive than that of the finest carriage owned by Louis XIV, who was, after all, an omnipotent king. And even Napoleon never had as much energy at his disposal as does a simple citizen of any OECD member country today.

We have access to a phenomenal amount of "abstract" energy, which people are not really aware of. The energy it takes to power a 100-watt light bulb can be generated by riding a bicycle at moderate speed. Now, imagine how much energy you would have to produce to supply a 900-watt toaster! What about a refrigerator, an entire household, or a car?

Considering the construction and insulation methods of the time, a house built in the 1950s consumed a fair amount of energy. While houses built since 2000 have benefitted from a host of technical innovations and vastly improved insulation, they are also a lot bigger, and their contact surface with the outside air has increased so much that, in the end, the unfortunate reality is that they consume as much energy as before.

The same goes for cars. Motor energy consumption has not really changed, because we keep making ever more powerful cars. They can now go from 0 to 100 km/h in 5 seconds, or even 4.5 seconds. This does not really serve any purpose, but we do it anyway.

As for today's refrigerators, it is true that they are better—from an energy standpoint—than those of yesteryear, but their lifespan is much shorter than it used to be. They are also bigger, and even with all our advanced technology, we have not been able to reduce the consumption of energy linked to refrigeration. And what about households that have two fridges, or a fridge and a freezer?

In each case, we have gained in comfort, certainly, but we are consuming more energy. And if the West has reduced its overall consumption by a tiny fraction in the last few years, other regions of the globe have steadily, and even outrageously, increased theirs. Technology in itself cannot solve our real problems; it tends rather to aggravate them. The predicted end of our petroleum reserves, and the questions of security surrounding our future supply of energy, force us to think, or at least they should. We should be thinking about the way we live, about what we really need to be happy, about our need to spend so much, go so fast, attend so many conferences—and always travel to them by plane!

Our entire economy is dependent on energy, and we are heading for a shortage. When that happens, when there is a disruption in supply, the people who asked the right questions beforehand may be able to offer some solutions. That said, one way or another, people are going to have to find a way to adapt. Humanity is about to get hit with some major changes. The suburbs, transportation, our high level of comfort... everything is going to change. We are not going to be able to heat our homes and workplaces in the same way that we do now. Woollen sweaters will be back in fashion when it gets cold!

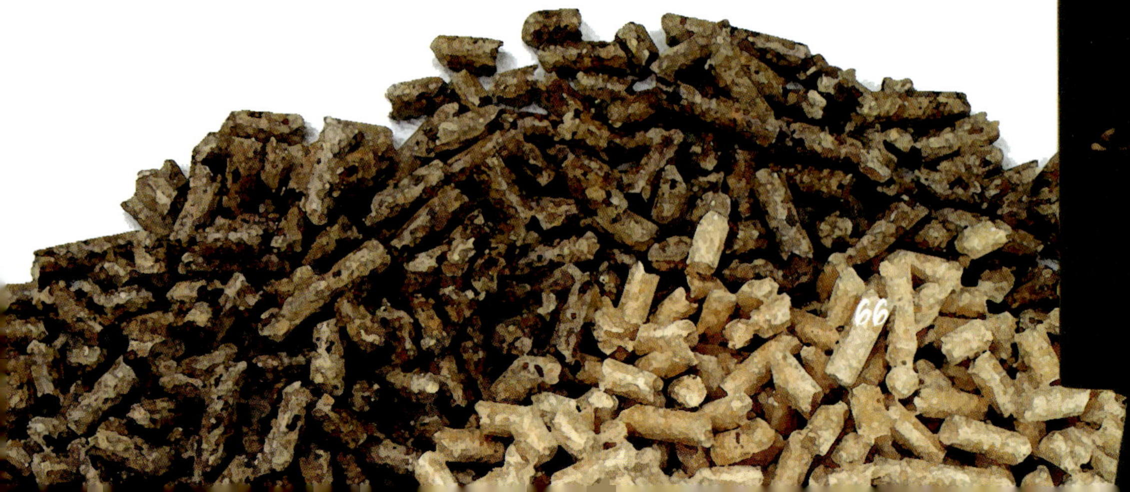

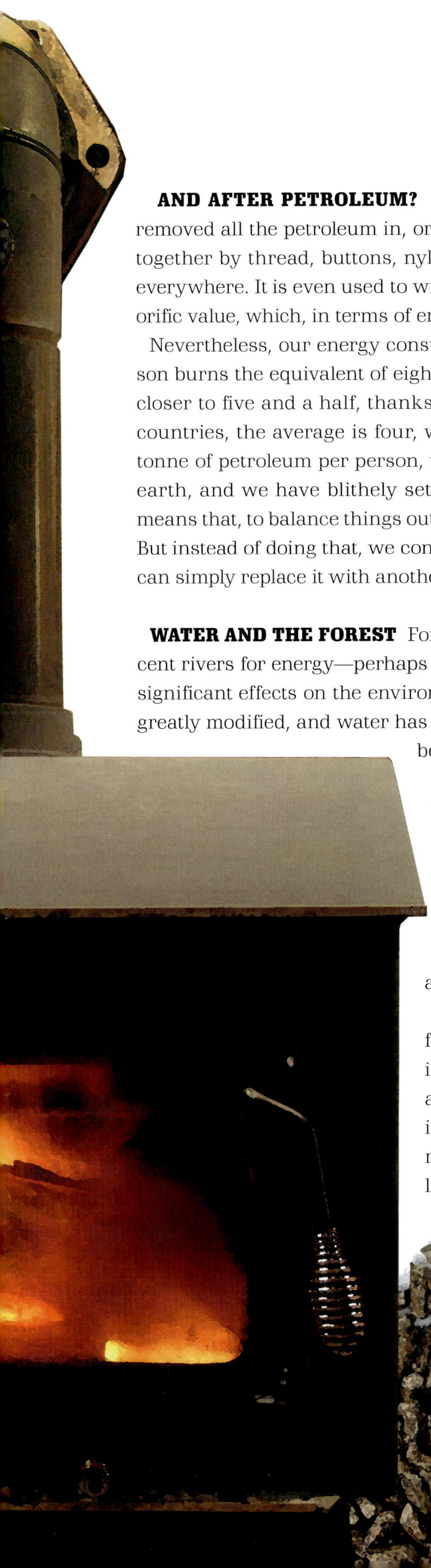

AND AFTER PETROLEUM? Petroleum is actually much too precious to burn! If we removed all the petroleum in, or on, our clothes, we would be... naked! Everything is held together by thread, buttons, nylon zippers, polyester... in short, petroleum. Petroleum is everywhere. It is even used to wrap our food. The problem is that it also has enormous calorific value, which, in terms of energy per unit of volume, or unit of mass, is unparalleled.

Nevertheless, our energy consumption per capita is staggering. In Canada, a single person burns the equivalent of eight and a half tonnes of petroleum annually. In Quebec, it is closer to five and a half, thanks to our hydropower resources. In all the OECD member countries, the average is four, while Asia comes in at 0.6. Even at the equivalent of one tonne of petroleum per person, we would have trouble meeting the needs of everyone on earth, and we have blithely set our course at somewhere between five and eight! This means that, to balance things out, we would have to consume five to eight times less energy. But instead of doing that, we continue believing that when one energy source runs out, we can simply replace it with another. The question is: which one?

WATER AND THE FOREST For many years, we have been harnessing our most magnificent rivers for energy—perhaps comprising their natural state—and our actions have had significant effects on the environment. The chemistry of the soil and the forests has been greatly modified, and water has become more acidic. Over time, a whole new balance has been established.

Each new project is inevitably located further away than the last. And the further we go, the more the costs of production and transportation of electricity increase. We built major energy-producing facilities because the cost per capita was worth it. But if we want to keep increasing our energy capacity, given that we have already travelled very far, we will have to come up with other exploitation strategies for our forests and watercourses.

The good news is that there is a lot of energy left in the forests and in the water. If we have to cool or heat a building that is close to a lake or a river, instead of installing a geothermal system, which is very costly, we could immerse a heat exchanger in the water and draw or reject as much energy as we need. The temperature of the lake or river would not be affected at all,

unless, of course, you were talking about a very large building and a very small lake or river. In thermal energy applications, water makes an excellent battery. In concrete terms, if we have to maintain the temperature inside a building at 20°C, whether the outside temperature is +40°C or −40°C, the water found in nature offers us an excellent solution. It is an incommensurable source of energy.

Wherever possible, we should envisage solutions like this to heat our communities and municipalities, rather than depending on petroleum from the Gulf of Mexico, or natural gas from Qatar. In a village surrounded by forest, a biomass boiler would be enough to heat the church, the rectory, the community centre, City Hall, and three or

four low-cost housing units. Not only does it make more sense, it is an opportunity worth exploiting.

We could increase points of local production. Some cities could even operate a number of small hydro power stations. This would no doubt have an impact on the resources, the fauna and flora, and the environment, but it would be minimal… you never get something for nothing. The important thing is to properly evaluate each initiative. That is sustainable development.

Whatever happens, there is no shortage of energy. As long as the sun shines, things grow, and as long as things grow, there will be available energy. In the past we used wood. Then we discovered coal, which is more concentrated, so less was needed to obtain the same amount of energy. Later we discovered petroleum, which offered an even better yield. But we are running out of it. Fortunately we still have wood. It has less calorific value, of course, but it does have the advantage of being renewable, provided we take care of it by giving it enough time to regenerate and do not abuse it.

TOWARD A SOCIAL PACT Whatever decisions we will have to make as a society in the near future, we have to motivate people to open their minds to the notion that we can do things differently. The first thing we have to do is to discuss ideas and offer solutions, and only then conduct environmental evaluations of projects that are retained. In that order! We often do things backwards, starting by ensuring a project's profitability, evaluating its technically feasibility, and then doing an environmental study to demonstrate its minimal impact. Then, at the end of the process, we approach people to try to sell the idea. That is a serious mistake.

The forest will always be there

Patrice, talk to us about the future of the forest...

THE FOREST IS OUR FUTURE The forest is the mother of us all. It is the matrix of all our futures; it is even the guarantee of humanity's future.

How can I be so certain?

Beyond international politics, which assures the stability of relations between peoples, societies, and countries, we urgently need to find fresh water supply solutions for agriculture and for human consumption. We have to make sure that we have enough energy and

natural resources to allow us to maintain sustainable socio-economic development, while controlling and, ideally, reducing global warming.

Let us use our imaginations and project ourselves a few decades into the future, to envisage the largely plausible scenario of the impact the forest might have on our, and its own, future.

A STOREHOUSE OF WATER Satisfying the world's need for water has become a gargantuan challenge. Products now not only leave a carbon footprint, but also a water footprint.

The efficient use of water has become as crucial as the efficient use of energy, because water is scarce. Desalinating sea water may seem like a viable solution on a local level, but it requires colossal amounts of energy. By drawing on the forest, we could, starting today, improve our global water balance, provided we respect water's natural cycle.

Today we know that forests support and replenish rivers. In the water cycle, we only borrow water on its way to the oceans. Because the forest acts like a sponge, it creates new resources by storing, beneath its surface, water from the oceans that has been evaporated by the sun; it increases the retention of water masses, and nourishes aquifer basins and ground water tables. After clearing land for centuries to expand agricultural land, a new concept of agriculture will no doubt lead humanity to alternate between fields and forests in ways that optimize their water resources and the way they use this 'blue gold.'

Of course, some forests are better suited than others to fulfill this role. But because of their natural potential to act as reservoirs, forests are inextricably tied to the solution to the question of global water supply.

MATERIALS OF THE FUTURE

Wood Since the dawn of humanity, wood has provided us with heat, shelter, and tools. We have found a thousand and one applications for it, whether for construction, furniture, printing, packaging, or energy production. Wood extracts are used as additives in many products, including foods! And what will become of wood tomorrow?

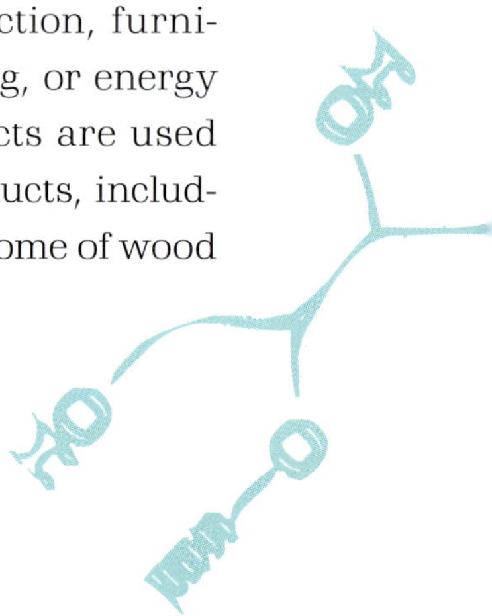

Tomorrow, wood will regain its magic. We already know that its use will increase in residential, commercial, and industrial construction. But that is not all. Wood is essentially composed of lignin—a substance that gives a tree trunk its strength—and sugars—more precisely cellulose, hemicellulose, and nanocellulose—along with a host of other components, especially lignans and antioxidants, a source of joy and wealth for the pharmaceutical companies. The knots in wood, for example, contain a powerful anticarcinogenic molecule, which is used in the treatment of breast cancer.

Lignin Present-day knowledge about the structure of lignin suggests that, in about ten years, the chemistry of lignin may gradually begin to replace the chemistry of petroleum. That means that all petroleum—and therefore nonrenewable—products will eventually be replaced by forest—or renewable—products. This shift in the chemistry of consumer products will probably take between 25 and 50 years to achieve. That may seem long compared to the lifespan of a human being, but it is very short in terms of society.

Since wood is a renewable material with a carbon-neutral cycle in terms of greenhouse gasses, its beneficial impact on climate change, and therefore on the future of humanity, will be considerable. As with water, we will simply be borrowing something from the forest: carbon dioxide will remain in our lignin-based products, for as long as we use them and until we return them to nature. On the other hand, in the case of petroleum (or coal), we release carbon dioxide into the air that had been stored for millennia!

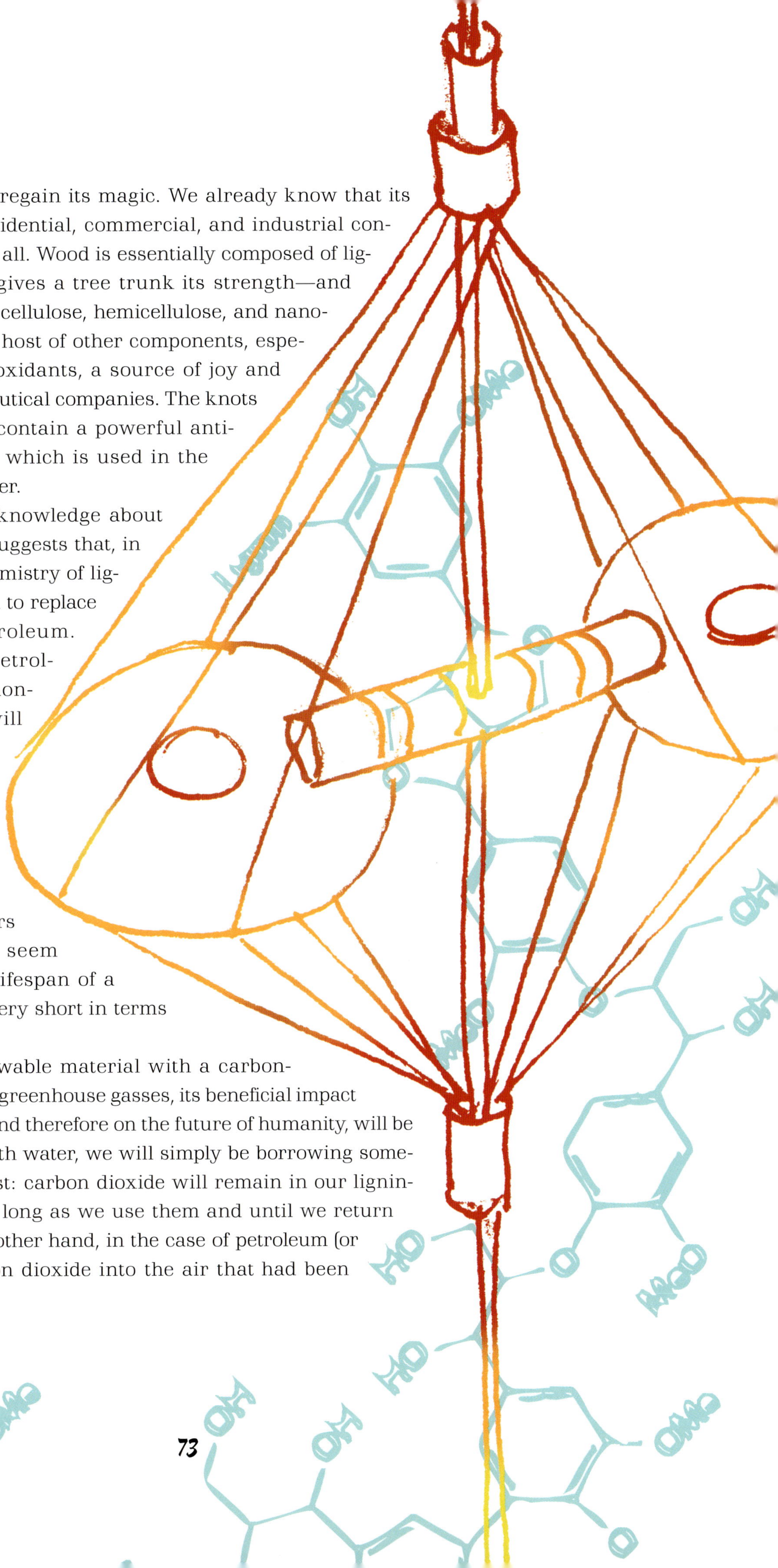

Sugars Enzymes modified by humans can transform the hemicellulose in wood into ethanol, a fuel that is already being added in various proportions to the gas we use to power our motor vehicles. Ethanol from the forest is no different from the ethanol that is obtained from fermenting starches, such as corn, for example, but it has the advantage of not having any negative impacts on the food chain, unlike corn, which requires enormous amounts of water to cultivate.

In addition to all the extraordinary properties being discovered by researchers week after week, nanocellulose has been found to be extremely strong. Its mechanical resistance is at least 10 times that of stainless steel, and probably close to one third that of a carbon nanotube—the strongest material we know. It easy to understand how it could be used in the fabrication of lightweight composite materials for the aeronautic and automobile industries. I can already imagine a paper airplane propelled by solar energy, or an elevator cable capable of launching payloads into low earth orbit. Are you smiling? Well, the sceptics will be proven wrong!

Like conventional petroleum refineries, we are now developing forest biorefineries, where we will soon be separating wood extractives—bark, lignin, cellulose—into their constituent molecules, to be able to reconstitute them into products that are useful for society. Let there be no doubt: these new products are the materials of the future, because they come from abundant and renewable forest resources. Eventually, we will destructure wood into basic elements to reconstruct it according to our needs. We are learning from nature and the forest.

FOREST BIOENERGY The production of energy, and especially of biofuels from forest biomass, will allow us to come full circle while reconnecting the bond of forest and humanity. Early man used energy locally when he burned wood to stay warm. Today, wood pellets can be used both for residential heating and for producing larger amounts of energy. And while they are a "civilized" version of the wood of our ancestors, wood pellets do not necessarily constitute the only, or the best, use of the forest biomass. In a post-petroleum era, our future depends on the development of biofuels from the forest with more widespread applications. We will soon be able to manufacture biodiesel fuels from forest biomass. In the near future, the forest will make a significant contribution to the reduction of the effects of the world's vehicle stock on global warming.

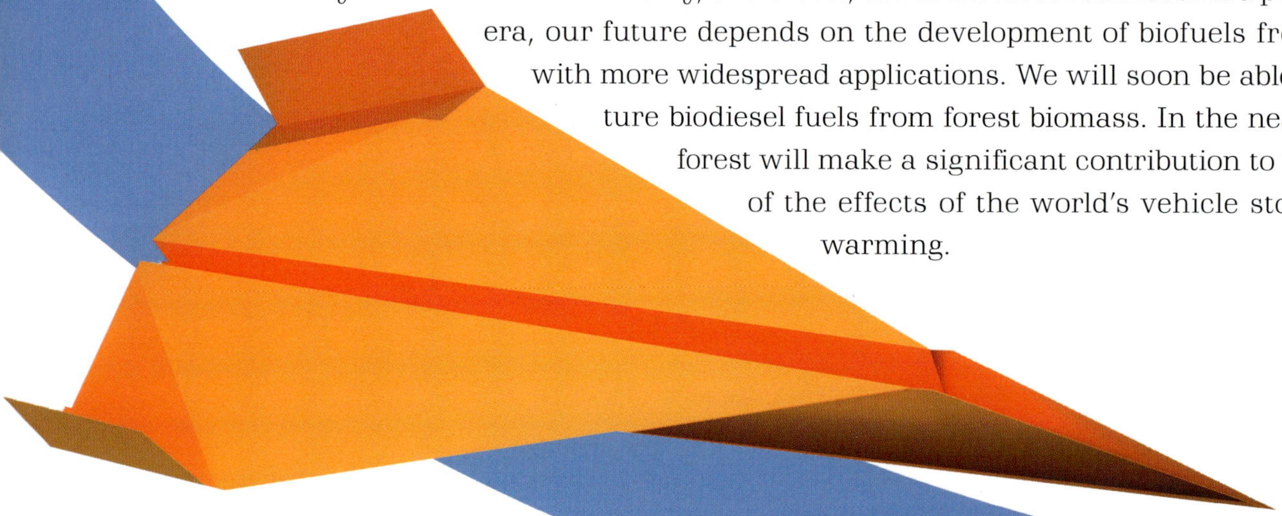

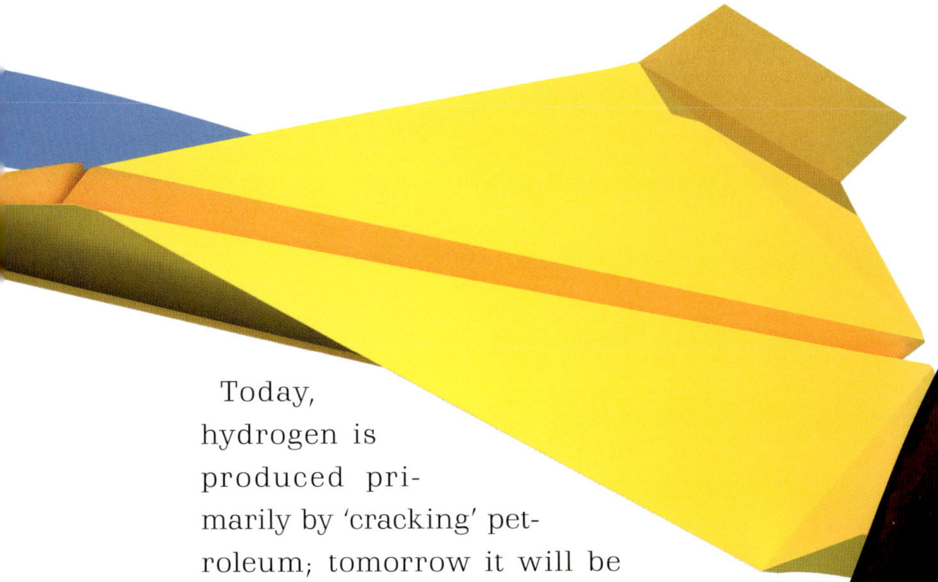

Today, hydrogen is produced primarily by 'cracking' petroleum; tomorrow it will be made from the forest biomass. Technical advances on the hydrogen-powered engine and hydrogen fuel cells have made it possible to build a hydrogen-fuelled car. Advancements on the storing and distribution of hydrogen will lead to its general use as a fuel, using the same distribution networks as diesel or gasoline.

Whatever the world's population, its needs will not be satisfied without the forest. The forest is humanity's future, and it deserves to be recognized and valued. Is it not true that knowledge overcomes prejudice? So let us continue our quest for education and knowledge. The only limit to humankind's future is our imagination. The forest teaches us this.

The experts

LANDSCAPE. Philippe Poullaouec-Gonidec Philippe Poullaouec-Gonidec is a professor at the School of Architecture and Landscape of the Université de Montréal. He also holds the UNESCO Chair in Landscape and Environmental Design (CUPEUM) and the Chair for Landscape and Environment from the same institution. A Fellow of the Trudeau Research Foundation, he has been directing research projects for many years on the characterization of landscapes and the process of their social and cultural invention. He was knighted by the Ordre des Arts et des Lettres (Order of Arts and Letters) (France), an honour that underscored his contribution to the cultural recognition of landscape and to the contemporary art of the garden. The co-founder of the International Garden Festival at the Reford Gardens in Grand-Métis (Quebec-Canada), he vacations in the Bas-Saint-Laurent region and is especially sensitive to marine and agroforest landscapes in this part of Quebec.
http://www.unesco-paysage.umontreal.ca

CLIMATE. Dominique Berteaux Dominique Berteaux is a research professor at Université du Québec à Rimouski, where he holds the Canada Research Chair in Northern Ecosystem Conservation. He is also the Director of the Groupe de recherche sur les environnements nordiques BORÉAS. His research on animal ecology has taken him to the heart of the great Canadian forests, from the Yukon to Quebec. For the last few years, he has been coordinating a major research project on the effects of climate change on biodiversity in Quebec.
http://chairedb.uqar.ca/welcome.html

IMAGINATION. Marie-Andrée Tardif Marie-Andrée Tardif is a painter. Her work is a song of praise to colour. Through rich tonality and vibrant contrasts that flow from an incandescent imagination, she exalts the beauty of nature. Her expressive canvasses translate reality and transform her wooded muse into a personal, symbolic, and poetic universe.
http://www.marieandreetardif.com

BIODIVERSITY. Daniel Gagnon Daniel Gagnon is a professor of Plant Ecology in the Department of Biological Sciences at Université du Québec à Montréal (UQAM). He is a member of the Centre for Forest Research and was the first research director at the Montréal Biodôme. He is also a founding trustee of the Eastern Townships Forest Research Trust, dedicated to the conservation and restoration of forests in southern Quebec and to the sustainable development of their resources. He teaches Biodiversity Conservation at the undergraduate and graduate level at UQAM.
http://www.cef-cfr.ca

ABORIGINAL PEOPLE. Konrad Sioui Konrad Sioui is the Grand Chief of the Huron-Wendat Nation. He describes himself as a chief dedicated to the service of his community and his Nation. He is actively engaged in their advancement and the defence of their rights.
http://www.wendake.ca

MEDICINE. Anny Schneider Anny Schneider is a writer, herbalist, and naturopath. She is widely recognized as an expert in the identification of the wild medicinal flora of Quebec, a province through which she has travelled extensively every summer for the last twenty years. A fascinating and passionate lecturer, she is also an entertainer, a poet, and a polemist, dedicated to the environment and social justice. She writes for a variety of local and corporate publications and can also be seen on *repere.tv*.
http://www.annyschneider.com

LEISURE. Sylvie Miaux Sylvie Miaux is a professor in the Department of Leisure, Culture and Tourism Studies of the Université du Québec à Trois-Rivières. She is especially interested in the experience of walking. Since she was a child, the forest has been a special place for her, somewhere where she could escape, awaken her senses, and let her imagination run free. Walking in the forest remains her favourite year-round leisure activity. Alone or accompanied, Sylvie strolls along forest paths looking for mushrooms, berries, and unexpected discoveries. For her, leisure in the forest is one of the best ways to reconnect with nature and oneself.
http://www.uqtr.ca

ECOTOURISM. Julianna Priskin Julianna Priskin is as passionate about exploring Canada's forests as she is about preserving their ecological integrity. A citizen of the world who practices ecological responsibility, she lived and worked in Europe and Australia before moving to Canada. Her research is primarily dedicated to better understanding how to manage tourism and recreation, with the goal of minimizing their effects on the environment and transforming them into tools for conservation while helping to solve some of the issues facing the planet today. Recently, Julianna worked in the forests of Quebec, with Parks Canada, to evaluate the nature of the visitor experience in protected areas. She participates in various ongoing projects as a research consultant in sustainable tourism development, and is also an associate professor in the Department of Urban Studies and Tourism at the Université du Québec à Montréal.
http://www.uqam.ca

WOOD. Hervé Deschênes Hervé Deschênes is the Vice-President of FPInnovations. He has also held positions as Director of International Projects and Manager of the Wood Transformation Group at TECSULT Inc.; Director, Wood Transformation Unit, Centre d'enseignement et de recherche en foresterie de Sainte-Foy inc. (CERFO); Director, Wood Engineering Section, École normale supérieure d'enseignement technique, Sofati training (Cameroon); Coordinator, Department of Forest Technology, CÉGEP Sainte-Foy; and Assistant Superintendent of Supplies for two Quebec mills belonging to the Domtar forestry corporation. He also served as President and General Manager of the Association de santé et sécurité des pâtes et papiers du Québec inc. and the Association de santé et sécurité des industries de la forêt du Québec inc. He holds a bachelor's degree in Forest Science from Université Laval and a graduate diploma in Business Administration.
http://www.fpinnovations.ca

PAPER. Jean Hamel Jean Hamel has been working for FPInnovations, Canada's main forest products research institute, for 25 years. As Vice-President of the Pulp and Paper division, he is involved in managing research projects, technical services, and technology transfers in the areas of wood harvesting, wood products, pulp and paper, nanotechnology, bioenergy generation, and chemical production. Jean obtained a bachelor's degree in Mechanical Engineering in 1983, and a master's degree in the same discipline in 1985, both from Université de Sherbrooke, in Quebec. He discovered his passion for papermaking as a student while participating in a co-op program with paper mills located on the North Shore of Quebec and in the Trois-Rivières region. There, he became aware of the full potential of the art of papermaking, which unites centuries-old knowledge with ultramodern technologies.
http://www.fpinnovations.ca

ENERGY. Daniel Rousse Daniel Rousse holds the Chair for Industrial Research in Energy Technology and Energy Efficiency at the École de technologie supérieure. He obtained a bachelor's and a master's degree from École Polytechnique de Montréal, as well as a PhD from McGill University. He has done research for Canadair, Canada Post Corporation, Doppelmayr Ski Lifts Company, CanmetENERGY, and ACT Informatique. He taught at École Polytechnique de Montréal and Université Laval, held the position of Vice-Rector at Université du Québec à Rimouski, before moving on to become the Director of Network Development at Université du Québec. He is dedicated to fostering respect for both the social and natural environment.
http://www.t3e.info

FUTURE. Patrice Mangin Patrice J. Mangin is the General Director of the Centre intégré en pâtes et papiers and a professor of Chemical Engineering at Université du Québec in Trois-Rivières. He was the CEO of the Centre technique du papier (Grenoble, France), served as the Stora-Enso Chair for the Royal Institute of Technology (Stockholm), and was a researcher at the Pulp and Paper Research Institute of Canada. As well, he has presided over a number of organizations, including the TAPPI Research Management Committee (United States), the Confederation of European Paper Industries Research Group (Brussels), and the Technical Association for the Graphic Arts (TAGA, United States). He is currently the President of the Canadian Pulp and Paper Network for Innovation in Education and Research (PAPIER), a member of Agenda 2020 (Washington), and an executive member of the Pulp and Paper Technical Association of Canada (PAPTAC). Passionate about paper, printing, and the forest, and a pragmatic visionary, Patrice is widely recognized as a champion of international cooperation.
http://www.cipp.ca

This work was printed on Supreme gloss paper,
and set in Schadow BT typeface,
Design and layout by Manon E. Léveillé.
Printed on November 29, 2010
by Transcontinental, Sherbrooke, Canada,
for IQ Press.